10 Gerogia Milestones Grade 8 Math Practice Tests

The Ultimate Test Prep Collection with Answer Explanations

Dr. A. Nazari

10 Practice Tests

 *Grade 8 Mathematics* 

Welcome!

*This book contains **10 full-length practice tests** — the most comprehensive preparation you can get for your Grade 8 math assessment. Each test covers all six topics:*

📖 *Irrational Numbers* 📖 *Powers & Scientific Notation*

📖 *Linear Equations* 📖 *Functions*

📖 *Geometry* 📖 *Data & Relationships*

Ten tests give you the practice needed to walk into the real test feeling fully prepared.

Thorough preparation leads to outstanding results.

❝ Ten full tests! By the time you finish, there won't be any surprises on test day. ❞

 How to Use This Book

A complete 10-test preparation program

- **10 Full-Length Practice Tests** — each covers all 6 chapters of Grade 8 math: irrational numbers, exponents & scientific notation, linear equations, functions, geometry, and data analysis.

- **Detailed Answer Explanations** — every question includes a step-by-step solution so you learn from every mistake.

- **Formula Reference Sheet** — all the key Grade 8 formulas you need, organized and ready for quick review.

- **Test Tracker** — log your scores across all 10 tests and monitor your progress from start to finish.

⭐ PHASE 1: Foundation (Tests 1–3)

Untimed or soft-timed. Focus on understanding the format, identifying strengths and weaknesses, and building good study habits.

⭐⭐ PHASE 2: Building Skills (Tests 4–7)

Timed (70 minutes each). Work on pacing, accuracy, and showing complete solutions. Review weak topics between tests.

⭐⭐⭐ PHASE 3: Test-Day Ready (Tests 8–10)

Full test conditions: strict timing, quiet space, no notes. Compare scores with your early tests to see your growth.

Schedule: Take one test every 3–4 days, or one per week. Use study days between tests to review.

Types of Questions

 **Multiple Choice:** *Four options — work the problem first, then match. Eliminate obviously wrong answers to narrow your choices.*

Short Answer & Constructed Response: *Show every step: equations, substitutions, simplifications. Partial credit rewards correct reasoning even if the final answer is off.*

Graphing & Data Analysis: *Plot points, draw lines, interpret graphs. Label axes clearly.*

Tip: Ten tests is a full preparation program. Don't rush. The key is what you do between tests — study, review, and understand your mistakes before moving forward.

Find more at
ViewMath.com/GA-Grade8

💡 Test-Taking Tips 💡

Your complete test-day toolkit

🕐 Before the Test

- Review your notes from the previous test — focus on your weak topics
- Set up a quiet, clean workspace with all your materials ready
- Start with a positive mindset: you've prepared for this

✏️ During the Test

- Read each problem fully before calculating anything
- Write the formula or set up the equation first, then substitute values
- Show all your work — every step, every operation
- If stuck for more than 2 minutes, mark it and move on
- Use estimation to check if your answers are reasonable

After the Test

- Read the full explanation for every question you got wrong
- Write down which topics gave you trouble (not just question numbers)
- Study those topics before taking the next test
- Record your score in the Test Tracker

⚠️ Common Mistakes in Grade 8 Math

⚠️ **Exponents:** $(ab)^n = a^n b^n$, but $a^m + a^n \neq a^{m+n}$. Only multiply/divide to combine.

⚠️ **Slope formula:** $m = \frac{y_2 - y_1}{x_2 - x_1}$ — keep the order consistent.

⚠️ **Systems of equations:** The solution must satisfy both equations.

⚠️ **Transformations:** Rotations and reflections change position; dilations change size.

⚠️ **Volume:** Use $\pi \approx 3.14$ or leave as π — match what the question asks.

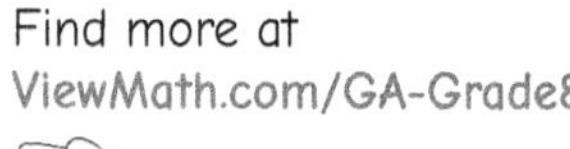

Find more at
ViewMath.com/GA-Grade8

🧰 What You'll Need 🧰

Gather these materials before you begin

Sharpened Pencils — #2 pencils, at least two

Good Eraser — for clean corrections

Scratch Paper — for working out problems

Ruler / Straightedge — for graphing & geometry

Quiet Space — no distractions

Focused Mind — ready to do your best

- ✔ Pencils and eraser
- ✔ Scratch paper (provided on official test day)
- ✔ Ruler or straightedge (if required)
- ✔ Protractor (if required)

- ✖ Calculator (unless your state test allows it)
- ✖ Cell phone or any electronic device
- ✖ Notes, textbooks, or reference sheets
- ✖ Help from others during the test

♥ A Note for Parents & Guardians

Ten tests is a comprehensive program. Plan **one test every 3–4 days** (or one per week) with study sessions between each test.

How to help:

- *Tests 1–3 should be untimed — build understanding before adding pressure.*
- *After each test, review the answer explanations together. Ask: "Which topics were hardest? Let's study those before the next one."*
- *Use the Test Tracker to celebrate progress over time.*
- *For topic-specific help, pair this book with our **Grade 8 Math Study Guide** or **Grade 8 Workbook**.*

Find more at
ViewMath.com/GA-Grade8

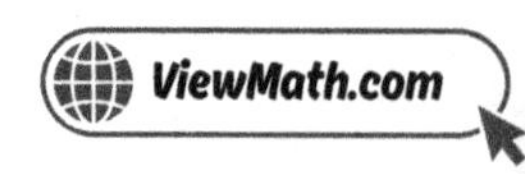

Grade 8 Formula Reference

Keep this page handy — you may use it during your practice tests!

X^1 Exponent Rules

$$a^m \cdot a^n = a^{m+n} \qquad (a^m)^n = a^{mn} \qquad (ab)^n = a^n \cdot b^n$$

$$\frac{a^m}{a^n} = a^{m-n} \qquad a^0 = 1 \; (a \neq 0) \qquad a^{-n} = \frac{1}{a^n}$$

Lines & Linear Equations

Slope: $m = \dfrac{y_2 - y_1}{x_2 - x_1} = \dfrac{rise}{run}$

m = slope   b = y-intercept

Slope-intercept: $y = mx + b$

Parallel lines: same slope

Proportional: $y = mx$

Proportional: passes through origin

Scientific Notation

$a \times 10^n \quad$ where $1 \leq |a| < 10$

Multiply: *add exponents*

Divide: *subtract exponents*

$\sqrt{x}$ Roots & Number Sense

Perfect squares: 1, 4, 9, 16, 25, 36, 49, 64, 81, 100, 121, 144

Perfect cubes: 1, 8, 27, 64, 125

$\sqrt{2} \approx 1.414 \qquad \sqrt{3} \approx 1.732 \qquad \pi \approx 3.14159$

Pythagorean Theorem & Distance

$a^2 + b^2 = c^2 \qquad c$ = hypotenuse (longest side of a right triangle)

Distance: $d = \sqrt{(x_2 - x_1)^2 + (y_2 - y_1)^2}$

Volume Formulas

Cylinder $V = \pi r^2 h \qquad$ **Cone** $V = \dfrac{1}{3}\pi r^2 h \qquad$ **Sphere** $V = \dfrac{4}{3}\pi r^3$

⌇ Angle Relationships

Triangle angle sum: 180°   **Exterior angle** = sum of two remote interior angles

Parallel lines + transversal: Alternate interior angles are equal · Co-interior angles sum to 180°

⬒ Functions

Each input → exactly one output   **Vertical line test:** if any vertical line hits graph more than once ⇒ not a function

Linear: constant rate of change ($y = mx + b$)   **Nonlinear:** rate of change varies

↻ Transformations

Translation: slide   **Reflection:** flip   **Rotation:** turn   **Dilation:** resize

Congruent = same shape & size   *Similar = same shape, proportional size*

Tip: *Bookmark this page! Review it before each test so these formulas become second nature.*

Find more at
ViewMath.com/GA-Grade8

Multiplication Table

×	1	2	3	4	5	6	7	8	9	10	11	12
1	1	2	3	4	5	6	7	8	9	10	11	12
2	2	4	6	8	10	12	14	16	18	20	22	24
3	3	6	9	12	15	18	21	24	27	30	33	36
4	4	8	12	16	20	24	28	32	36	40	44	48
5	5	10	15	20	25	30	35	40	45	50	55	60
6	6	12	18	24	30	36	42	48	54	60	66	72
7	7	14	21	28	35	42	49	56	63	70	77	84
8	8	16	24	32	40	48	56	64	72	80	88	96
9	9	18	27	36	45	54	63	72	81	90	99	108
10	10	20	30	40	50	60	70	80	90	100	110	120
11	11	22	33	44	55	66	77	88	99	110	121	132
12	12	24	36	48	60	72	84	96	108	120	132	144

💡 How to Use This Table

To find **4 × 7**:

1. Find **4** in the left column (blue).
2. Find **7** in the top row (blue).
3. Follow the row and column until they meet: the answer is **28**!

> ⓘ **Tip:** You can also use this table for division! If you know $28 \div 4 =$?, find 28 in the 4's row. The column header gives you the answer: **7!**

📈 My Test Tracker 📈

Name: ________________________________   **Start Date:** ____________________

GETTING STARTED (Tests 1–3)

Test 1 — Untimed

Date: ____________   Score: _______ / _______   %: _______   Topics to review: ____________________

Test 2 — Untimed

Date: ____________   Score: _______ / _______   %: _______   Topics to review: ____________________

Test 3 — Soft Timer

Date: ____________   Score: _______ / _______   %: _______   Topics to review: ____________________

BUILDING SKILLS (Tests 4–7)

Test 4 — Timed (70 min)

Date: ____________   Score: _______ / _______   %: _______   Focus area: ____________________

Test 5 — Timed (70 min)

Date: ____________   Score: _______ / _______   %: _______   Focus area: ____________________

Test 6 — Timed (70 min)

Date: ____________   Score: _______ / _______   %: _______   Focus area: ____________________

Test 7 — Timed (70 min)

Date: ____________   Score: _______ / _______   %: _______   Focus area: ____________________

TEST-DAY READY (Tests 8–10)

Date: ___________   Score: ______ / ______   %: ______   Growth since Test 1: ___________________

Date: ___________   Score: ______ / ______   %: ______   Growth since Test 1: ___________________

Date: ___________   Score: ______ / ______   %: ______   Growth since Test 1: ___________________

Score Progress

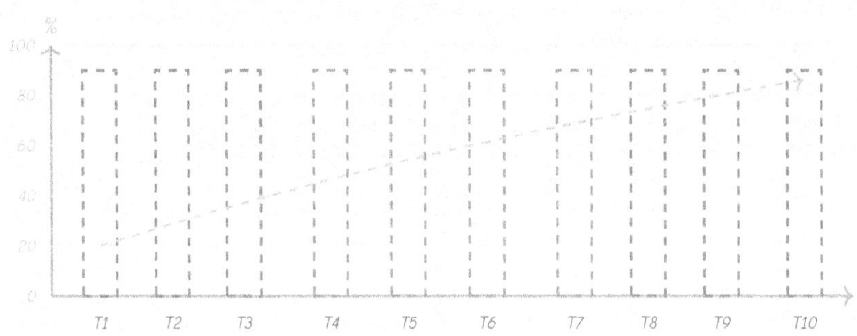

Shade each bar after every test. Watch your improvement!

Final Reflection

The most important thing I learned: ____

The topic where I improved the most: ____

My advice for other students: ____

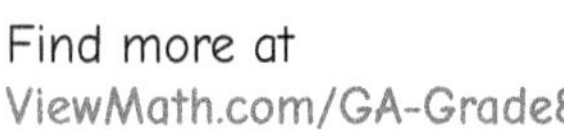
Find more at
ViewMath.com/GA-Grade8

⭐ Table of Contents ⭐

Here's what we'll explore together!

 Let's learn and have fun!

Practice Test 1

☑ 30 Questions

✏ Before You Start ✏

- ✓ **Read each question carefully** before choosing your answer.
- ✓ **Show your work** on scratch paper when you need to.
- ✓ **Skip hard questions** and come back to them later.
- ✓ **Check your answers** when you're done.
- ✓ **Take your time** — there's no rush!

⭐ You've Got This! ⭐

Do your best and show what you know!

1. *True or false: The number $\frac{22}{7}$ is equal to π.*

Your Answer

2. *The table below shows four repeating decimals and the power of 10 a student used to convert each. Which one uses the WRONG power of 10?*

Decimal	Multiply by
$0.\overline{7}$	10
$0.\overline{45}$	100
$0.\overline{123}$	100
$0.\overline{8}$	10

(A) $0.\overline{7}$

(B) $0.\overline{45}$

(C) $0.\overline{123}$

(D) $0.\overline{8}$

3. *The diagram below shows a rectangle with irrational side lengths. Which is the best estimate of the area?*

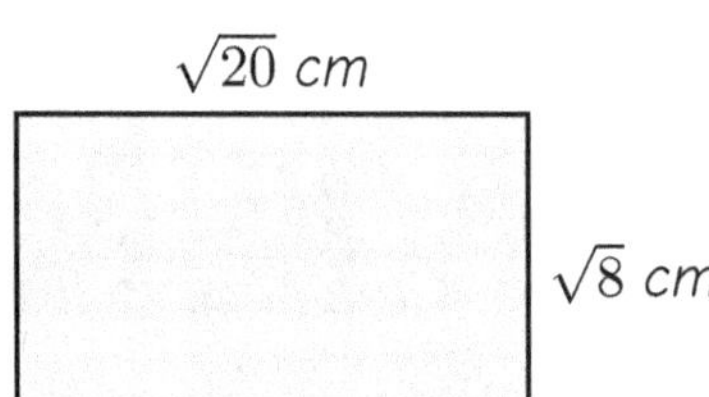

(A) $\sqrt{28} \approx 5.3\ cm^2$

(B) $\sqrt{160} \approx 12.6\ cm^2$

(C) $28\ cm^2$

(D) $160\ cm^2$

4. *Simplify* $7^5 \cdot 7^{-5}$.

(A) 7^{25}

(B) 7^{10}

(C) 0

(D) 1

5. *The diagram shows a square and a cube. The square has an area of A square units and the cube has a volume of V cubic units. If $A = V$, which pair of side lengths is correct?*

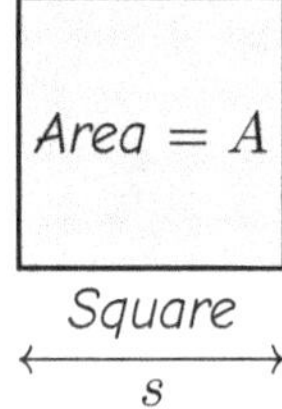

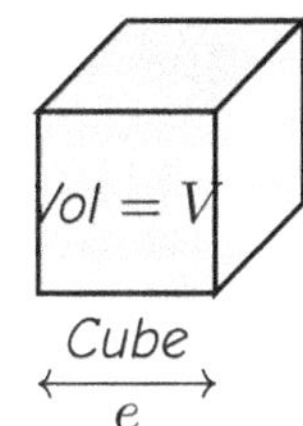

(A) $s = 6,\ e = 6$

(B) $s = 8,\ e = 4$

(C) $s = 9,\ e = 3$

(D) $s = 5,\ e = 5$

6. *Estimate the product* $498{,}000{,}000 \times 6{,}200$ *by first writing each factor in scientific notation and then computing.*

Your Answer:

7. Study the diagram below. Each box produces its output by multiplying the two inputs. What belongs in the output box marked "?"?

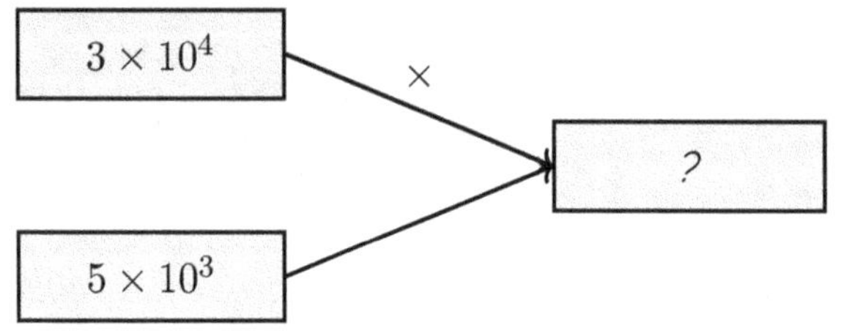

A 8×10^7

B 1.5×10^7

C 1.5×10^8

D 15×10^7

8. A bike rental company charges \$8 per hour. Which equation models the total cost y for x hours?

A $y = x + 8$

B $y = 8x$

C $y = \frac{x}{8}$

D $y = 8x + 10$

9. Find the slope through $(3, -1)$ and $(7, 11)$.

Your Answer:

10. Solve: $x + y = 15$ and $x - y = 3$.

Your Answer:

11. The graph shows two lines representing the earnings of two workers. Worker A earns a fixed amount plus an hourly rate, and Worker B earns only an hourly rate.

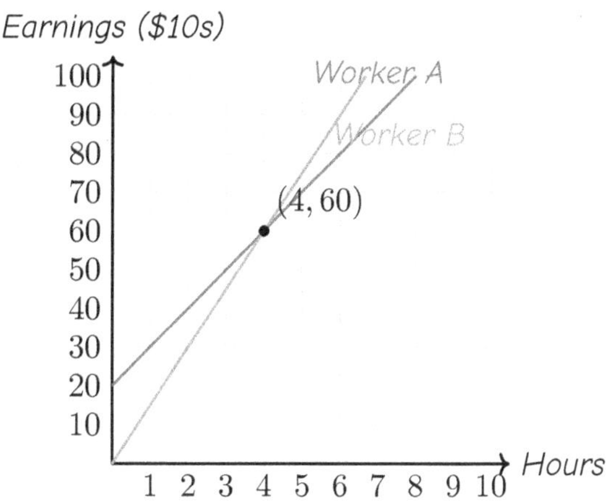

After 4 hours, both workers have earned $60. If Worker B works 6 hours, how much more does Worker B earn than Worker A?

A) $5

B) $10

C) $15

D) $20

12. Look at the graph of the relation below. Explain whether it is a function and why.

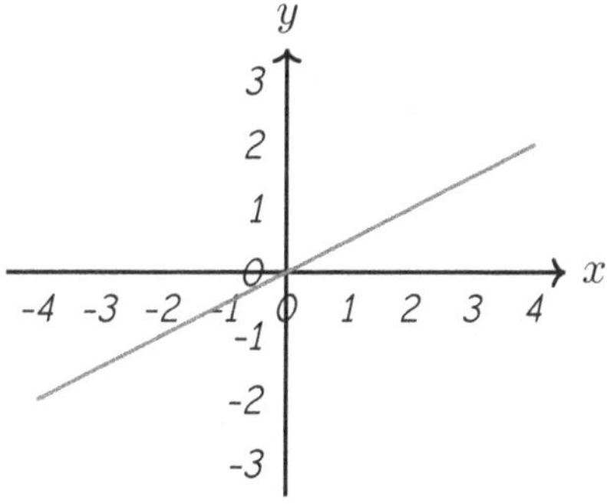

Your Answer:

Find more at
ViewMath.com/GA-Grade8

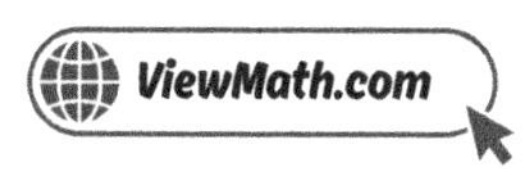

13. If $f(x) = \frac{x+4}{2}$, what is $f(6)$?

Your Answer:

14. Function A: $y = 5x + 3$. Function B: $y = 2x + 10$. Which function has a greater rate of change?

(A) Function A

(B) Function B

(C) They have the same rate of change.

(D) Cannot be determined.

15. Give an example of a nonlinear equation.

Your Answer:

16. A pool has 150 gallons and is filling at 10 gallons per minute. What is the function?

(A) $y = 150x + 10$

(B) $y = 10x + 150$

(C) $y = 160x$

(D) $y = 10x - 150$

17. A student fills a glass with water, drinks half, then refills it. Describe the graph of water level vs. time.

Your Answer:

18. Point $N(0, 4)$ is rotated $180°$ around the origin. What are the coordinates of N'?

Your Answer:

Find more at
ViewMath.com/GA-Grade8

19. *Which statement is true about all congruent figures?*

(A) *They are always in the same orientation.*

(B) *They are always in the same quadrant.*

(C) *One can always be mapped onto the other by rigid transformations.*

(D) *They must be mirror images of each other.*

20. *A triangle with vertices $(0,0)$, $(6,0)$, and $(6,8)$ is dilated by factor $\frac{1}{2}$ from the origin. What is the perimeter of the image?*

(A) 24

(B) 12

(C) 6

(D) 48

21. *Triangle PQR has sides 5, 12, 13. After a dilation by $k = 2$, what is the longest side of the image?*

(A) 13

(B) 15

(C) 24

(D) 26

22. *Which angles are equal when parallel lines are cut by a transversal?*

(A) *Co-interior angles*

(B) *Supplementary angles*

(C) *Corresponding angles*

(D) *Adjacent angles*

23. *Is a triangle with sides 7, 24, 25 a right triangle?*

(A) *Yes, because $7 + 24 > 25$.*

(B) *Yes, because $7^2 + 24^2 = 25^2$.*

(C) *No, because $7^2 + 24^2 \neq 25^2$.*

(D) *No, because all sides must be equal.*

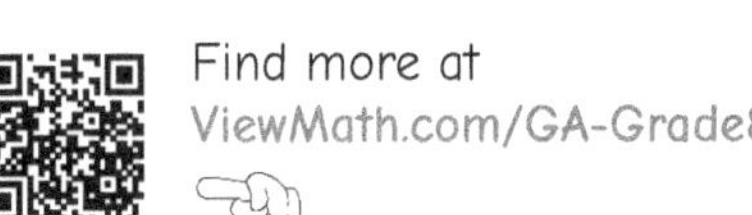
Find more at
ViewMath.com/GA-Grade8

24. Which expression represents the distance between (x_1, y_1) and (x_2, y_2)?

(A) $(x_2 - x_1) + (y_2 - y_1)$

(B) $\sqrt{(x_2 - x_1)^2 + (y_2 - y_1)^2}$

(C) $(x_2 - x_1)^2 + (y_2 - y_1)^2$

(D) $|x_2 - x_1| \times |y_2 - y_1|$

25. A cylindrical water tank has radius 5 ft and height 12 ft. How much water can it hold? Use $\pi \approx 3.14$.

(A) $942\ ft^3$

(B) $188.4\ ft^3$

(C) $300\ ft^3$

(D) $471\ ft^3$

26. What type of association does this scatter plot show?

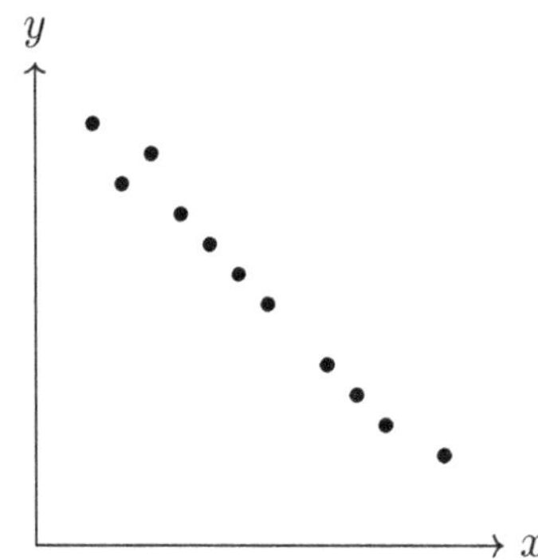

(A) Positive linear

(B) Negative linear

(C) No association

(D) Nonlinear

27. A scatter plot has 10 data points. One outlier is far above the trend. When drawing the line of best fit, you should:

(A) draw the line through the outlier

(B) ignore the outlier and fit the line to the other 9 points

(C) not draw a line at all

(D) draw the line only through the outlier

28. A trend line gives $y = -3x + 45$. Data was collected for $x = 1$ to $x = 10$. Predict y when $x = 5$ and when $x = 30$. State which prediction is more reliable and why.

Your Answer

29. Using the table above, what percentage of 7th graders ride a bike?

(A) 55%

(B) 45%

(C) 62.5%

(D) 37.5%

30. Data: $\{2, 2, 2, 2, 12\}$. Calculate the MAD and explain the effect of the outlier.

Your Answer

 *End of Practice Test 1* 

Great job finishing the test!

☑ My Score

I got ______________ out of 30 questions right.

*Check your answers in the **Answer Key** at the back of the book.*

💡 Review any questions you missed. That's how we learn!

📊 Check Your Score Online!

Visit **ViewMath Academy** to enter your answers and see which topics you need to review. You can also explore lessons, take quizzes, track your scores, and save your progress!

viewmath.com/score/8.1.GA.16

Or go to viewmath.com/score and enter code: 8.1.GA.16

2

Practice Test 2

30 Questions

✏️ Before You Start ✏️

- ✓ **Read each question carefully** before choosing your answer.
- ✓ **Show your work** on scratch paper when you need to.
- ✓ **Skip hard questions** and come back to them later.
- ✓ **Check your answers** when you're done.
- ✓ **Take your time** — there's no rush!

⭐ You've Got This! ⭐

Do your best and show what you know!

1. The number line below shows two points, P and Q.

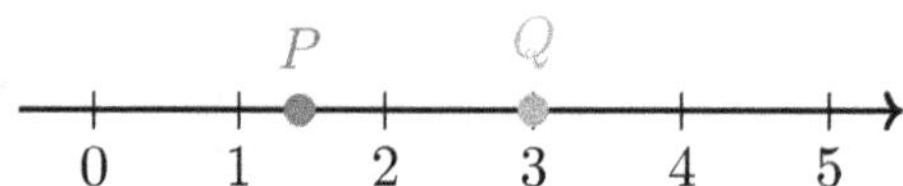

Point P represents $\sqrt{2}$ and point Q represents 3. Which statement is true?

(A) Both P and Q are rational.

(B) Both P and Q are irrational.

(C) P is irrational and Q is rational.

(D) P is rational and Q is irrational.

2. The flowchart below shows the conversion steps for a repeating decimal. Fill in the missing values to convert $0.\overline{54}$ to a fraction.

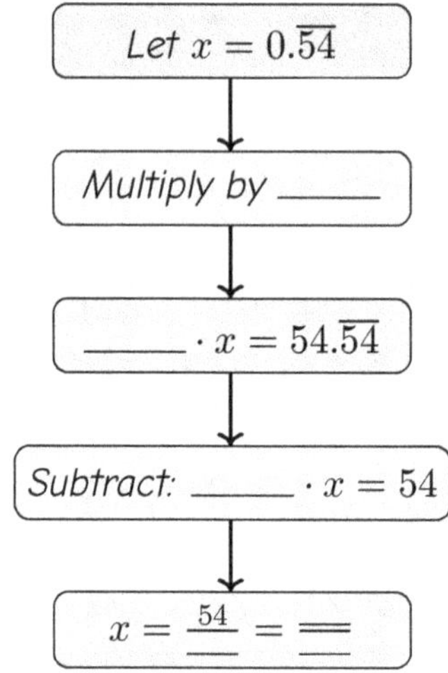

Your Answer

3. The number line below shows the values of π and $\sqrt{10}$ marked. Find the distance between them.

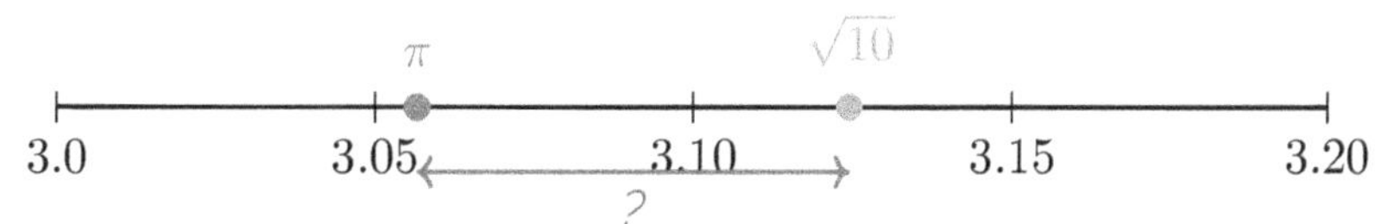

Your Answer:

4. Which expression is equivalent to 4^{-3}?

(A) -64

(B) -12

(C) $\frac{1}{12}$

(D) $\frac{1}{64}$

5. Solve $x^2 = \frac{9}{16}$.

(A) $x = \frac{3}{4}$ only

(B) $x = \pm\frac{3}{4}$

(C) $x = \frac{9}{8}$

(D) $x = \pm\frac{9}{4}$

6. Which of the following is true about 4.2×10^{-3} and 4.2×10^3?

(A) They are equal

(B) 4.2×10^3 is 10^6 times as large as 4.2×10^{-3}

(C) 4.2×10^{-3} is 10^6 times as large as 4.2×10^3

(D) Their product is 1

7. What is $\frac{8 \times 10^7}{2 \times 10^3}$?

(A) 4×10^4

(B) 6×10^4

(C) 4×10^{10}

(D) 16×10^{10}

Find more at
ViewMath.com/GA-Grade8

ViewMath.com

8. A faucet leaks at a constant rate. It leaks 45 mL in 15 minutes. How many mL does it leak in 1 hour?

Your Answer:

9. A plant grows from 4 cm to 16 cm in 6 weeks. What is the rate of change?

(A) 1 cm/week

(B) 2 cm/week

(C) 3 cm/week

(D) 4 cm/week

10. Solve: $3x + y = 10$ and $y = x - 2$.

Your Answer:

11. Apples cost \$1.50 each and oranges cost \$1.00 each. You buy 12 fruits for \$14.50. How many apples did you buy?

(A) 3

(B) 4

(C) 5

(D) 6

12. Give an example of a set of three ordered pairs that is a function.

Your Answer:

Find more at
ViewMath.com/GA-Grade8

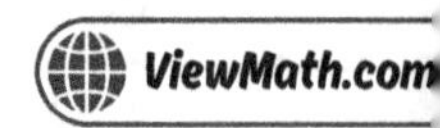

13. *The graph of h is shown below. For what value of x does $h(x) = 6$?*

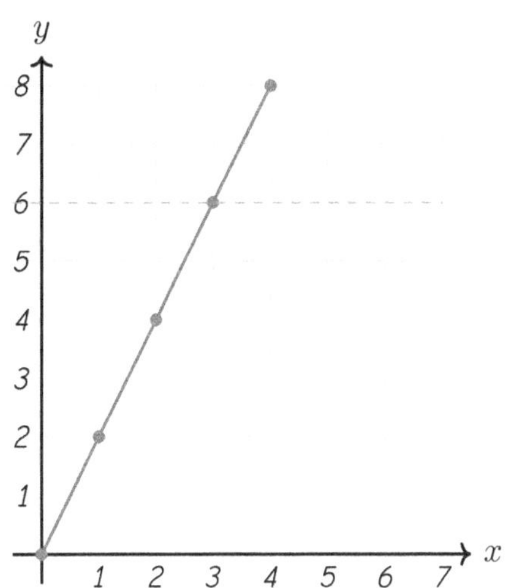

Your Answer:

14. *Company A charges $20 plus $8 per hour. Company B charges $50 plus $3 per hour. After how many hours do they charge the same amount?*

Your Answer:

15. *The graph of $y = 7$ is:*

(A) *a vertical line*

(B) *a horizontal line*

(C) *a parabola*

(D) *a curve*

16. *The function $y = 50x + 200$ models a bank account balance after x weeks. What is the weekly deposit?*

(A) $50

(B) $100

(C) $200

(D) $250

17. A bathtub fills up, then someone soaks, then the water drains. Which describes the water level graph?

(A) Decreasing, constant, increasing

(B) Increasing, decreasing, constant

(C) Increasing, constant, decreasing

(D) Constant, increasing, decreasing

18. After a translation, a triangle's longest side was originally 8 cm. What is the longest side of the image?

(A) 4 cm

(B) 8 cm

(C) 16 cm

(D) It depends on the direction of the translation.

19. Two congruent rectangles each have a perimeter of 28 cm. One has a length of 9 cm. What is its width?

Your Answer:

20. Point $(-3, 5)$ is rotated 90° counterclockwise around the origin. What is its image?

(A) $(5, 3)$

(B) $(-5, -3)$

(C) $(3, -5)$

(D) $(-5, 3)$

21. Two similar triangles have a scale factor of 4 from the smaller to the larger. If a side of the smaller triangle is 3 cm, what is the corresponding side of the larger triangle?

(A) 7 cm

(B) 0.75 cm

(C) 12 cm

(D) 9 cm

22. An equilateral triangle has all angles equal. What is each angle?

Your Answer:

Find more at
ViewMath.com/GA-Grade8

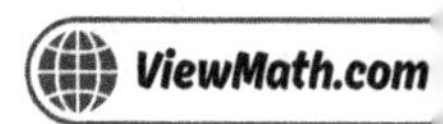

23. *A right triangle has legs 9 and 12. What is the hypotenuse?*

(A) 21

(B) 108

(C) 15

(D) $\sqrt{21}$

24. *Find the distance between $(-1, -1)$ and $(5, 7)$.*

Your Answer:

25. *What is the volume of the cone shown below in terms of π?*

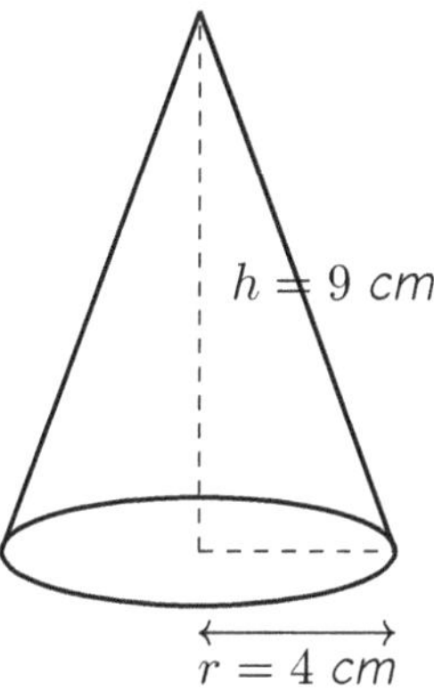

(A) $144\pi \ cm^3$

(B) $48\pi \ cm^3$

(C) $36\pi \ cm^3$

(D) $108\pi \ cm^3$

26. Describe the association shown in this scatter plot. Mention direction, shape, and any unusual features.

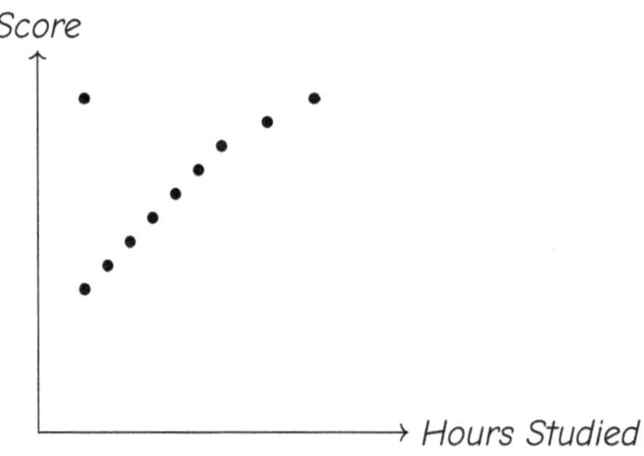

Your Answer:

27. What is a line of best fit?

(A) A line that passes through every data point

(B) A line that connects the first and last data points

(C) A straight line that comes close to most of the data points

(D) A vertical line through the middle of the scatter plot

28. Which statement about slope in a linear model is true?

(A) A positive slope means y decreases as x increases.

(B) A negative slope means y increases as x increases.

(C) A slope of zero means y stays constant.

(D) The slope is always positive in real data.

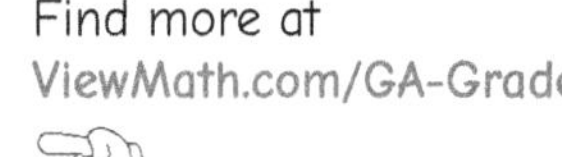

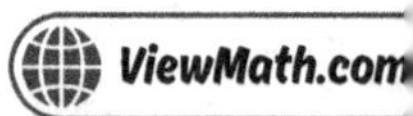

29. *A segmented (stacked) bar chart shows the proportion of students choosing band or chorus by gender. The boys' bar is 60% band and 40% chorus. The girls' bar is 45% band and 55% chorus. What can you conclude?*

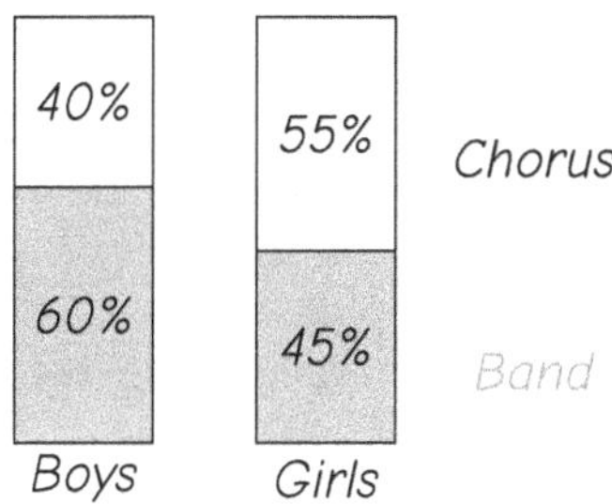

(A) There is no association between gender and music choice.

(B) Boys are more likely to choose band; girls more likely to choose chorus.

(C) Girls and boys prefer band equally.

(D) Chorus is more popular overall.

30. *A number line shows data points* $\{3, 5, 5, 7, 10\}$ *with the mean marked at 6. What is the MAD?*

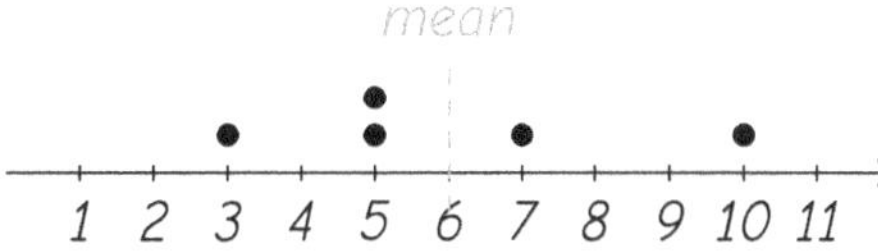

(A) 1.6

(B) 2

(C) 2.2

(D) 3

 # End of Practice Test 2 

Great job finishing the test!

 My Score

I got _____________ out of 30 questions right.

*Check your answers in the **Answer Key** at the back of the book.*

💡 *Review any questions you missed. That's how we learn!*

📊 Check Your Score Online!

Visit **ViewMath Academy** to enter your answers and see which topics you need to review. You can also explore lessons, take quizzes, track your scores, and save your progress!

viewmath.com/score/8.1.GA.17

Or go to viewmath.com/score and enter code: 8.1.GA.17

Practice Test 3

30 Questions

✏️ Before You Start ✏️

- ✓ **Read each question carefully** before choosing your answer.

- ✓ **Show your work** on scratch paper when you need to.

- ✓ **Skip hard questions** and come back to them later.

- ✓ **Check your answers** when you're done.

- ✓ **Take your time** — there's no rush!

⭐ You've Got This! ⭐

Do your best and show what you know!

1. Which number below can be written as a fraction $\frac{a}{b}$ where a and b are integers and $b \neq 0$?

   (A) $\sqrt{3}$

   (B) π

   (C) $0.\overline{81}$

   (D) $\sqrt{11}$

2. What is $0.\overline{27}$ as a fraction in simplest form?

   (A) $\frac{27}{100}$

   (B) $\frac{27}{99}$

   (C) $\frac{3}{11}$

   (D) $\frac{9}{33}$

3. Which is larger: $4\sqrt{2}$ or $\sqrt{30}$?

   (A) $4\sqrt{2}$, because $4 > \sqrt{30}$

   (B) $\sqrt{30}$, because $30 > 2$

   (C) $4\sqrt{2}$, because $4\sqrt{2} \approx 5.66 > 5.48 \approx \sqrt{30}$

   (D) They are equal.

4. Which of the following is equal to $\frac{8^6}{8^6}$?

   (A) 8^{36}

   (B) 8^0

   (C) 0

   (D) 8

5. A square has an area of 196 square centimeters. What is the length of one side?

   (A) 13 cm

   (B) 49 cm

   (C) 14 cm

   (D) 98 cm

6. Which of the following is 0.00072 written in scientific notation?

   (A) 7.2×10^{-4}

   (B) 7.2×10^4

   (C) 72×10^{-5}

   (D) 0.72×10^{-3}

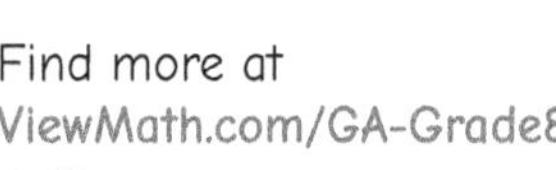
Find more at
ViewMath.com/GA-Grade8

7. A virus has a mass of 9.5×10^{-18} grams. If a sample contains 2×10^6 viruses, what is the total mass?

- (A) 1.9×10^{-11} g
- (B) 1.9×10^{-12} g
- (C) 11.5×10^{-12} g
- (D) 1.9×10^{-24} g

8. Store A sells apples for \$2 per pound. Store B's prices are shown in the table.

Pounds	3	6	9
Cost (\$)	7.50	15.00	22.50

Which store has the lower unit price?

- (A) Store A
- (B) Store B
- (C) They charge the same price.
- (D) Not enough information.

9. A car's odometer reads 120 miles at 2:00 PM and 270 miles at 5:00 PM. What is the car's speed (rate of change)?

- (A) 45 mph
- (B) 50 mph
- (C) 60 mph
- (D) 90 mph

10. How many solutions does this system have? $y = 2x + 5$ and $4x - 2y = -10$.

- (A) No solution
- (B) One solution
- (C) Two solutions
- (D) Infinitely many solutions

11. A snack bar sells hot dogs for \$3 and burgers for \$5. One day 80 items were sold for \$340. How many hot dogs were sold?

- (A) 20
- (B) 30
- (C) 40
- (D) 50

Find more at
ViewMath.com/GA-Grade8

12. The ordered pairs $(a, 5)$ and $(a, 12)$ belong to a relation. Can this relation be a function? Write Yes or No.

Your Answer

13. If $f(x) = 4x - 3$ and $f(x) = 17$, what is x?

(A) 3

(B) 4

(C) 5

(D) 7

14. Function P: $y = 3x + 12$. Function Q: $y = 7x + 12$. Which function grows faster? Write P or Q.

Your Answer

15. A table shows $(1, 5)$, $(2, 10)$, $(3, 15)$, $(4, 20)$. What is the rate of change, and is the function linear?

(A) Rate of change $= 5$; linear

(B) Rate of change $= 5$; nonlinear

(C) Rate of change varies; linear

(D) Rate of change varies; nonlinear

16. A table shows $(0, 8)$, $(3, 20)$, $(6, 32)$. What is the equation?

(A) $y = 3x + 8$

(B) $y = 4x + 8$

(C) $y = 8x + 4$

(D) $y = 6x + 8$

17. The graph below shows a function. How many sections does it have, and what are they?

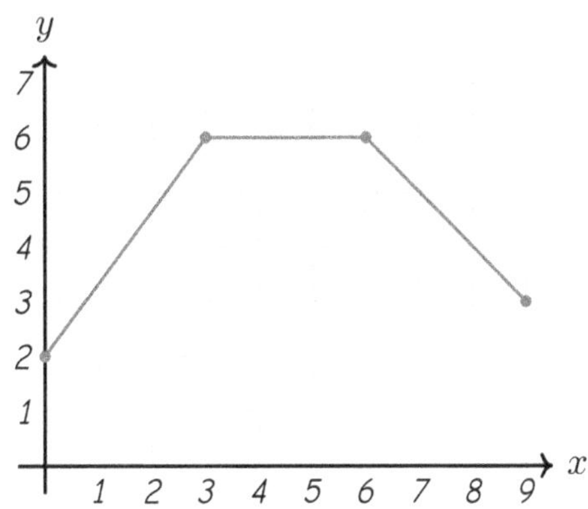

(A) 2 sections: increasing, decreasing

(B) 3 sections: increasing, constant, decreasing

(C) 3 sections: decreasing, constant, increasing

(D) 4 sections

18. Which of the following is a rigid transformation?

(A) Dilation by a factor of 2

(B) Stretching a figure horizontally

(C) Reflection over a line

(D) Shrinking a figure to half its size

19. Two figures have the same shape but one is twice as large. Are they congruent?

(A) Yes, same shape means congruent.

(B) Yes, because their angles match.

(C) No, they are similar but not congruent.

(D) No, they are neither similar nor congruent.

20. Which transformation maps (x, y) to $(-y, x)$?

(A) Reflection over the x-axis

(B) Reflection over the y-axis

(C) 90° counterclockwise rotation

(D) 180° rotation

21. Two similar rectangles have corresponding sides in the ratio $2 : 5$. If the shorter side of the smaller rectangle is 8 cm, what is the shorter side of the larger rectangle?

    (A) 3.2 cm

    (B) 13 cm

    (C) 16 cm

    (D) 20 cm

22. Find the value of x in the triangle below.

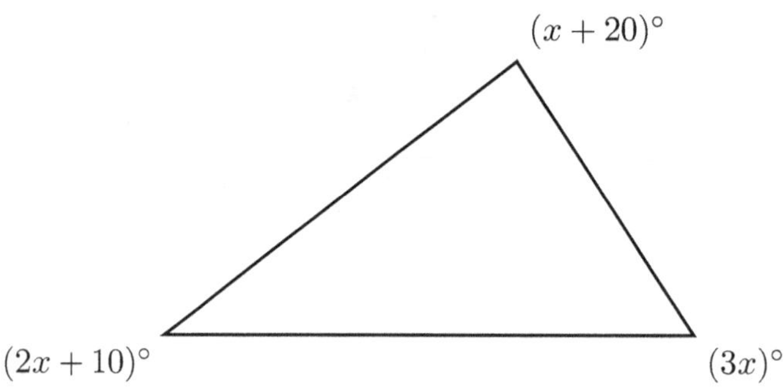

Your Answer

23. Is the triangle with sides 9, 40, 41 a right triangle?

    (A) No, because $9 + 40 \neq 41$.

    (B) No, because $9^2 + 40^2 \neq 41^2$.

    (C) Yes, because $9^2 + 40^2 = 41^2$.

    (D) Yes, because $41 - 40 = 1$.

24. What is the distance between $(-4, -3)$ and $(4, 3)$?

    (A) $\sqrt{14}$

    (B) 14

    (C) 10

    (D) $\sqrt{10}$

5. The volume of a sphere with radius r is $\frac{4}{3}\pi r^3$. If the radius is tripled, the new volume is:

(A) 3 times the original

(B) 9 times the original

(C) 27 times the original

(D) 81 times the original

6. Can a scatter plot show a positive association that is nonlinear? Give an example.

Your Answer

7. A scatter plot shows a strong nonlinear (curved) pattern. Should you draw a straight line of best fit?

(A) Yes — always draw a straight line.

(B) No — a straight line does not fit curved data well.

(C) Yes — but only if there are outliers.

(D) No — you should never draw any line.

8. Using the model $y = -4x + 100$, predict y when $x = 10$.

(A) 140

(B) 60

(C) 96

(D) 40

9.

	Dog	Cat	Total
Boys	15	10	25
Girls	12	13	25
Total	27	23	50

How many girls prefer dogs?

(A) 10

(B) 12

(C) 13

(D) 15

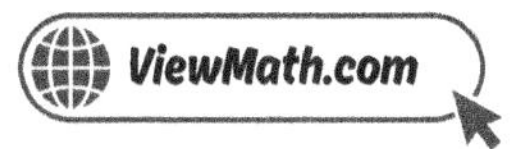

30. If every value in a data set is multiplied by 3, the MAD:

(A) stays the same

(B) is multiplied by 3

(C) is divided by 3

(D) is multiplied by 9

Find more at
ViewMath.com/GA-Grade8

 # End of Practice Test 3 

Great job finishing the test!

☑ My Score

I got __________ out of 30 questions right.

Check your answers in the **Answer Key** at the back of the book.

💡 Review any questions you missed. That's how we learn!

📊 Check Your Score Online!

Visit **ViewMath Academy** to enter your answers and see which topics you need to review. You can also explore lessons, take quizzes, track your scores, and save your progress!

viewmath.com/score/8.1.GA.18

Or go to viewmath.com/score and enter code: 8.1.GA.18

4

Practice Test 4

☑ 30 Questions

✏ Before You Start ✏

- ✔ **Read each question carefully** before choosing your answer.
- ✔ **Show your work** on scratch paper when you need to.
- ✔ **Skip hard questions** and come back to them later.
- ✔ **Check your answers** when you're done.
- ✔ **Take your time** — there's no rush!

⭐ You've Got This! ⭐

Do your best and show what you know!

1. Which of the following numbers is irrational?

   (A) $\frac{5}{8}$

   (B) $0.\overline{6}$

   (C) $\sqrt{9}$

   (D) $\sqrt{7}$

2. Write $0.5\overline{8}$ as a fraction in simplest form.

Your Answer

3. A student claims that $2\sqrt{5} = \sqrt{10}$. Is this correct?

   (A) Yes, because $2 \times 5 = 10$.

   (B) No, $2\sqrt{5} = \sqrt{20}$, not $\sqrt{10}$.

   (C) Yes, multiplication distributes into the square root.

   (D) No, $2\sqrt{5} = 4\sqrt{5}$.

4. Evaluate $5^0 + 5^{-1}$. Express your answer as a decimal.

Your Answer

5. What is $\sqrt{49}$?

   (A) 6

   (B) 7

   (C) 8

   (D) 24.5

6. What is 6.1×10^3 written in standard form?

   (A) 61,000

   (B) 610

   (C) 6,100

   (D) 0.0061

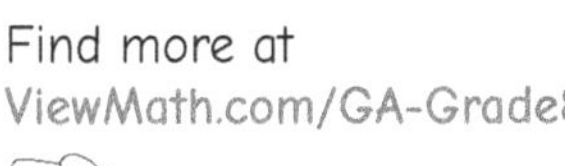
Find more at
ViewMath.com/GA-Grade8

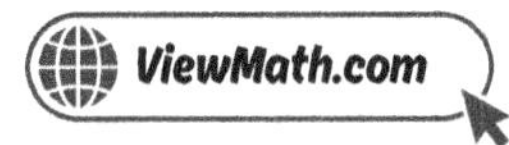

7. A bacterium has a length of 2×10^{-6} m. If 1 micrometer (μm) $= 10^{-6}$ m, what is the bacterium's length in micrometers?

(A) $2 \times 10^{-12} \; \mu m$

(B) $0.002 \; \mu m$

(C) $2 \; \mu m$

(D) $2{,}000 \; \mu m$

8. Taxi A charges \$3 per mile. Taxi B charges \$36 for a 9-mile trip. Which taxi has the higher unit rate?

Your Answer:

9. What is the slope through $(-4, -1)$ and $(2, 5)$?

(A) -1

(B) $\frac{2}{3}$

(C) 1

(D) $\frac{3}{2}$

10. Solve: $y = -x + 5$ and $y = 2x - 1$.

(A) $(1, 4)$

(B) $(2, 3)$

(C) $(3, 2)$

(D) $(4, 1)$

11. Two trains leave the same station heading in opposite directions. Train A goes 60 mph and Train B goes 80 mph. After how many hours are they 420 miles apart?

(A) 2

(B) 2.5

(C) 3

(D) 3.5

12. A relation contains the ordered pairs $(7, 3)$ and $(7, 9)$. Is this relation a function? Explain.

Your Answer:

13. *The table below shows values for $f(x)$. If $f(x) = 15$, what is x?*

x	0	1	2	3	4
$f(x)$	3	6	9	12	15

(A) 2

(B) 3

(C) 4

(D) 5

14. *A table for Function C shows $(0, 7)$, $(1, 11)$, $(2, 15)$. What is the initial value?*

15. *A table shows $(0, 3)$, $(1, 7)$, $(2, 11)$, $(3, 15)$. Is this linear or nonlinear?*

(A) *Nonlinear, because the outputs are odd numbers only.*

(B) *Linear, because the change in y is always 4.*

(C) *Nonlinear, because the outputs get larger.*

(D) *Linear, because the first output is positive.*

16. The graph shows a line passing through two labeled points. What is the slope?

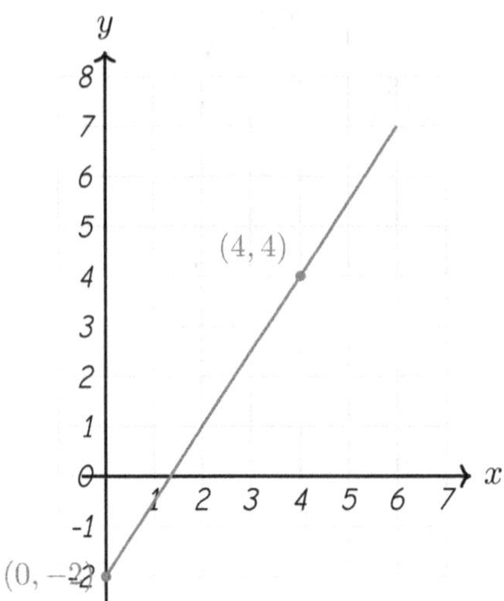

(A) $\frac{1}{2}$

(B) 1

(C) $\frac{3}{2}$

(D) 2

17. A graph of water in a tank is a straight line sloping downward. What is happening?

(A) Water is being added at a constant rate.

(B) Water is leaking out at a constant rate.

(C) The tank is full and not changing.

(D) Water is being added at an increasing rate.

18. A student says that a reflection changes the area of a figure. Is this correct?

(A) Yes, because the figure flips.

(B) Yes, because one dimension reverses.

(C) No, because reflections preserve all measure-ments.

(D) No, but reflections change the perimeter.

Find more at
ViewMath.com/GA-Grade8

ViewMath.com

19. Rectangle $ABCD$ has $AB = 6$ and $BC = 10$. Rectangle $WXYZ$ has $WX = 10$ and $XY = 6$. Are they congruent?

(A) No, because the side lengths are in different order.

(B) No, because $6 \neq 10$.

(C) Yes, because the sets of side lengths are the same.

(D) Cannot be determined.

20. Point $A(0, -5)$ is rotated $90°$ counterclockwise around the origin. Where does A' land?

(A) $(5, 0)$

(B) $(0, 5)$

(C) $(-5, 0)$

(D) $(0, -5)$

21. A rectangle is 3 cm by 5 cm. Another rectangle is 6 cm by 9 cm. Are they similar?

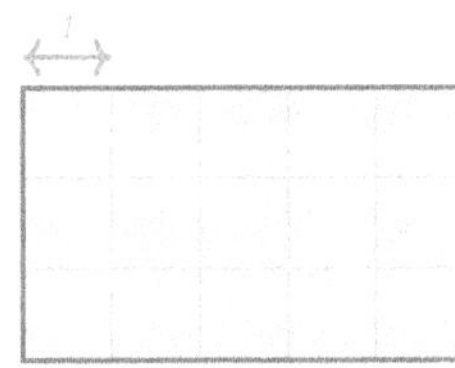

(A) Yes, with scale factor 2

(B) Yes, with scale factor 3

(C) No, the sides are not proportional.

(D) Yes, with scale factor $\frac{3}{5}$

22. An exterior angle of a triangle is $140°$. What is the adjacent interior angle?

(A) $40°$

(B) $140°$

(C) $50°$

(D) $220°$

Find more at
ViewMath.com/GA-Grade8

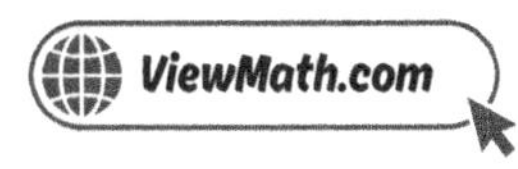

23. *A ladder leans against a wall as shown. How long is the ladder?*

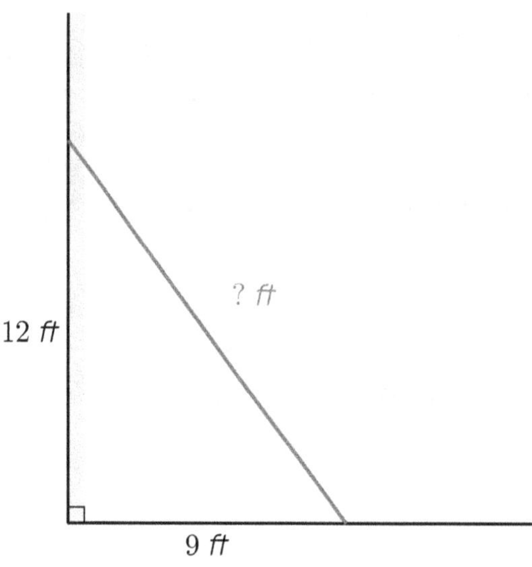

(A) 21 ft

(B) 3 ft

(C) 15 ft

(D) $\sqrt{21}$ ft

---

24. *Find the distance between $(-6, -8)$ and $(0, 0)$.*

Your Answer:

---

25. *What is the volume of a sphere with radius 3 cm? Use $\pi \approx 3.14$.*

(A) 28.26 cm^3

(B) 113.04 cm^3

(C) 36π cm^3

(D) 84.78 cm^3

---

26. *A scatter plot shows data points that form a curved (U-shaped) pattern. This is best described as:*

(A) positive linear association

(B) negative linear association

(C) nonlinear association

(D) no association

27. A trend line has equation $y = 1.5x + 2$. Using this, what is y when $x = 4$?

(A) 6

(B) 7

(C) 8

(D) 10

28. Data was collected for x-values from 1 to 10. Using the model to predict y when $x = 50$ is an example of:

(A) interpolation

(B) extrapolation

(C) correlation

(D) association

29. 80 students were surveyed about playing sports and playing a musical instrument.

	Plays sport	No sport	Total
Plays instrument	15	25	40
No instrument	30	10	40
Total	45	35	80

Is there an association between playing a sport and playing an instrument? Justify using conditional relative frequencies.

30. Data: $\{20, 25, 30, 35, 40\}$. Mean $= 30$. Which value has the greatest absolute deviation from the mean?

(A) 25

(B) 30

(C) 20

(D) 35

 # ⭐ *End of Practice Test 4* ⭐ 

Great job finishing the test!

 **My Score**

I got _____________ out of 30 questions right.

*Check your answers in the **Answer Key** at the back of the book.*

💡 Review any questions you missed. That's how we learn!

📊 **Check Your Score Online!**

*Visit **ViewMath Academy** to enter your answers and see which topics you need to review. You can also explore lessons, take quizzes, track your scores, and save your progress!*

viewmath.com/score/8.1.GA.19

Or go to viewmath.com/score and enter code: 8.1.GA.19

5

Practice Test 5

☑ *30 Questions*

✏ Before You Start ✏

✓ **Read each question carefully** before choosing your answer.

✓ **Show your work** on scratch paper when you need to.

✓ **Skip hard questions** and come back to them later.

✓ **Check your answers** when you're done.

✓ **Take your time** — there's no rush!

★ You've Got This! ★

Do your best and show what you know!

1. A calculator shows $\sqrt{144} = 12$. Is $\sqrt{144}$ rational or irrational? Explain.

2. To convert $0.\overline{63}$ to a fraction, which equation should you set up after letting $x = 0.\overline{63}$?

(A) $10x - x = 63$ (B) $100x - x = 63$

(C) $100x - x = 6.3$ (D) $1000x - x = 63$

3. If the side length of a square is $\sqrt{50}$ cm, what is the perimeter?

(A) $4\sqrt{50} \approx 28.3$ cm   (B) 50 cm

(C) $\sqrt{200} \approx 14.1$ cm   (D) 200 cm

4. Study the number line below. Which expression corresponds to the point marked with a star (★)?

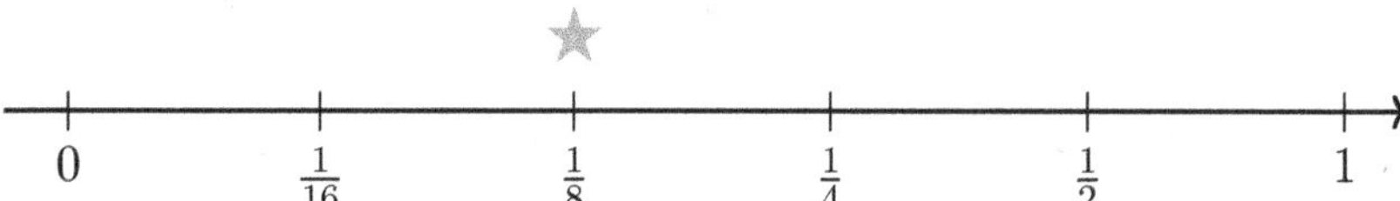

(A) 2^{-2} (B) 2^{-3}

(C) 2^{-4} (D) 2^{-1}

5. Which of the following is a perfect square?

(A) 50 (B) 72

(C) 81 (D) 90

6. *A hydrogen atom has a radius of about 2.5×10^{-11} meters. How many places do you move the decimal to write this in standard form?*

Your Answer:

7. *What is $5.2 \times 10^6 + 3.8 \times 10^6$?*

(A) 9×10^6

(B) 9×10^{12}

(C) 19.76×10^6

(D) 9×10^{36}

8. *Which ordered pair could NOT lie on the graph of a proportional relationship?*

(A) $(0,0)$

(B) $(1,5)$

(C) $(3,15)$

(D) $(2,13)$

9. *Find the slope through $\left(\frac{1}{2}, 3\right)$ and $\left(\frac{3}{2}, 7\right)$.*

(A) 2

(B) 4

(C) $\frac{1}{4}$

(D) 8

10. *Solve: $2x + 5y = 20$ and $2x + y = 8$.*

11. *Two numbers add to 50. Three times the smaller subtracted from twice the larger gives 30. Find both numbers.*

12. Which test is used to determine whether a graph represents a function?

   (A) Horizontal line test

   (B) Diagonal line test

   (C) Vertical line test

   (D) Slope test

13. A function table shows $f(1) = 4$, $f(2) = 7$, $f(3) = 10$. What is $f(2) - f(1)$?

Your Answer

14. Function R from the previous table has a rate of change of:

x	0	2	4	6
y	1	9	17	25

   (A) 2

   (B) 3

   (C) 4

   (D) 8

15. Which of the following is true about a nonlinear function?

   (A) Its graph is always a straight line.

   (B) Its rate of change is constant.

   (C) Its graph is curved.

   (D) It cannot have negative values.

16. A bike rental costs \$10 plus \$5 per hour. What is the total cost for 3 hours?

Your Answer

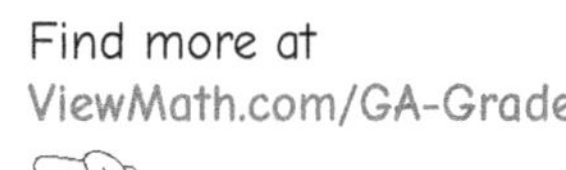

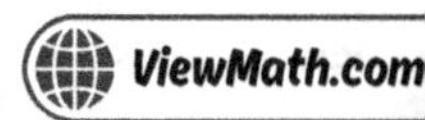

17. *A decreasing graph does NOT necessarily mean:*

(A) *The output values are getting smaller.*   (B) *The graph goes downward from left to right.*

(C) *The output values are negative.*   (D) *The rate of change is negative.*

18. *A triangle is rotated 90° clockwise around a point. Which property of the triangle changes?*

(A) *Side lengths*   (B) *Angle measures*

(C) *Position and orientation*   (D) *Perimeter*

19. Triangle RST has vertices $R(0,0)$, $S(4,0)$, and $T(0,3)$. Triangle $R'S'T'$ has vertices $R'(0,0)$, $S'(0,4)$, and $T'(-3,0)$. What transformation maps RST to $R'S'T'$?

(A) *Translation*   (B) *Reflection over the x-axis*

(C) *90° counterclockwise rotation*   (D) *90° clockwise rotation*

20. *Point (a,b) is rotated 180° and then reflected over the x-axis. The final image is:*

(A) (a,b) (B) $(-a,b)$

(C) $(a,-b)$ (D) $(-a,-b)$

21. Triangle A has sides 6, 8, 10. Triangle B has sides 9, 12, 15. What is the scale factor from A to B?

Your Answer:

22. *Two parallel lines are cut by a transversal. One angle measures 72°. What is its alternate interior angle?*

(A) 18°   (B) 72°

(C) 108°   (D) 288°

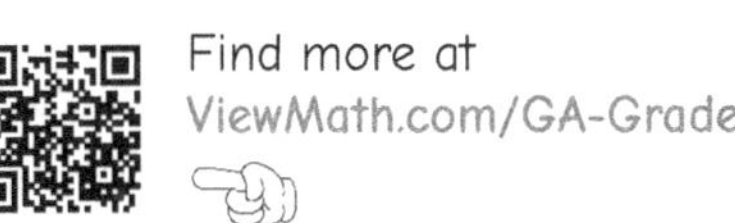

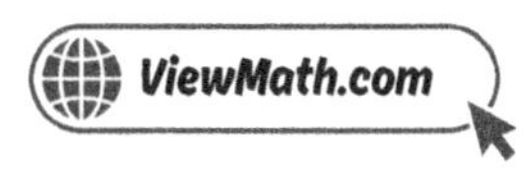

23. A right triangle has legs 8 and 15. What is the hypotenuse?

    (A) 23

    (B) 17

    (C) $\sqrt{23}$

    (D) 289

24. What is the distance between $(0, 0)$ and $(3, 4)$?

    (A) 7

    (B) 5

    (C) $\sqrt{7}$

    (D) 12

25. A cone has radius 6 cm and height 10 cm. What is its volume? Use $\pi \approx 3.14$.

    (A) 1130.4 cm^3

    (B) 376.8 cm^3

    (C) 188.4 cm^3

    (D) 565.2 cm^3

26. Which variable typically goes on the x-axis in a scatter plot?

    (A) The dependent variable

    (B) The response variable

    (C) The independent (explanatory) variable

    (D) The variable with the largest values

27. A trend line passes through $(2, 10)$ and $(8, 4)$. Find the slope and y-intercept.

    Your Answer:

28. Using the model $y = 4x + 7$, for what value of x is $y = 35$?

    (A) 7

    (B) 6

    (C) 8

    (D) 10

Find more at
ViewMath.com/GA-Grade8

29. Complete the two-way table and find the relative frequencies for each cell (as fractions of the total).

	Homework Yes	Homework No	Total
Grade A	24	6	
Not Grade A	16	14	
Total			

Your Answer

30. Two classes took the same test. Class X had MAD $= 5$ and Class Y had MAD $= 12$. Which class had more consistent scores?

(A) Class X

(B) Class Y

(C) Both were equally consistent

(D) Cannot be determined

Find more at
ViewMath.com/GA-Grade8

 # End of Practice Test 5 

Great job finishing the test!

☑ My Score

I got _____________ out of 30 questions right.

💡 *Review any questions you missed. That's how we learn!*

📊 Check Your Score Online!

Visit **ViewMath Academy** to enter your answers and see which topics you need to review. You can also explore lessons, take quizzes, track your scores, and save your progress!

viewmath.com/score/8.1.GA.20

Or go to viewmath.com/score and enter code: 8.1.GA.20

6

Practice Test 6

📋 30 Questions

✏️ Before You Start ✏️

- ✓ **Read each question carefully** before choosing your answer.
- ✓ **Show your work** on scratch paper when you need to.
- ✓ **Skip hard questions** and come back to them later.
- ✓ **Check your answers** when you're done.
- ✓ **Take your time** — there's no rush!

⭐ You've Got This! ⭐

Do your best and show what you know!

1. Is $\frac{5}{6}$ rational or irrational? Explain.

Your Answer:

2. When converting $0.\overline{123}$ to a fraction, by what power of 10 should you multiply both sides?

(A) 10

(B) 100

(C) 1000

(D) 10000

3. Order from least to greatest: π, $\sqrt{10}$, 3.2.

(A) π, 3.2, $\sqrt{10}$

(B) 3.2, π, $\sqrt{10}$

(C) $\sqrt{10}$, π, 3.2

(D) π, $\sqrt{10}$, 3.2

4. What is the value of $(-2)^0$?

(A) -2

(B) 0

(C) 1

(D) -1

5. A cube has a volume of 343 cubic inches. What is the length of one edge?

(A) 7 in.

(B) 49 in.

(C) $114.\overline{3}$ in.

(D) 17 in.

6. What is 9.03×10^{-5} written in standard form?

(A) 903,000

(B) 0.0000903

(C) 0.000903

(D) 0.00903

Find more at
ViewMath.com/GA-Grade8

7. A light year is approximately 9.5×10^{12} km. If a star is 4 light years away, how far is it in kilometers?

(A) 3.8×10^{13} km

(B) 3.8×10^{12} km

(C) 13.5×10^{12} km

(D) 9.5×10^{48} km

8. A graph shows a straight line through $(0,0)$ and $(2,9)$. What is the unit rate?

(A) 2

(B) 4.5

(C) 9

(D) 18

9. A line passes through $(2,7)$ and $(5,1)$. What is the slope?

(A) -2

(B) 2

(C) -3

(D) 3

10. On a graph, where is the solution to a system of two linear equations?

(A) At the y-intercept of the first line

(B) At the x-intercept of the second line

(C) At the point where the two lines intersect

(D) At the origin

11. Streaming Service A costs \$8 per month. Service B costs \$2 per month plus \$1.50 per movie. How many movies make Service B cost the same as Service A?

(A) 3

(B) 4

(C) 5

(D) 6

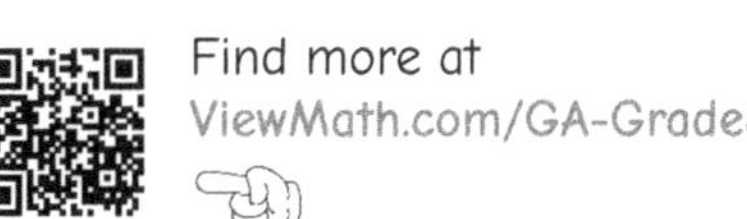

12. Look at the mapping diagram below. Does it represent a function?

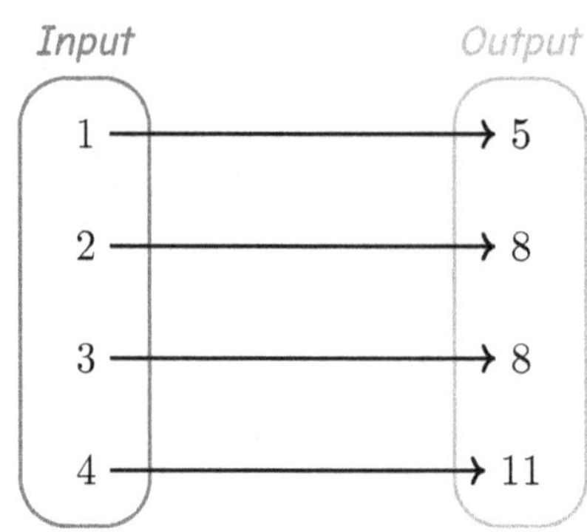

(A) No, because 8 appears twice as an output.

(B) No, because inputs and outputs are different numbers.

(C) Yes, because each input has exactly one arrow going out.

(D) Yes, because no output is negative.

13. If $f(x) = 4x + 1$, find the value of $f(3) + f(1)$.

Your Answer

14. Function A: $y = -x + 20$. Function B: $y = -3x + 20$. Which function decreases faster?

(A) Function A

(B) Function B

(C) They decrease at the same rate.

(D) Neither is decreasing.

15. The function $y = -2x + 6$ is:

(A) Nonlinear, because the slope is negative.

(B) Nonlinear, because y can be negative.

(C) Linear, because it has the form $y = mx + b$.

(D) Linear, because it passes through the origin.

16. A line passes through $(3, 7)$ and $(6, 16)$. What is the y-intercept?

(A) -2

(B) 0

(C) 1

(D) 3

17. Describe the graph of a cup of hot coffee cooling down to room temperature over an hour.

18. Point $K(-3, -5)$ is rotated $90°$ counterclockwise around the origin. What are the coordinates of K'?

Your Answer:

19. Triangle P has angles $40°$, $60°$, and $80°$. Triangle Q has angles $40°$, $60°$, and $80°$. Are they congruent?

(A) Yes, because matching angles guarantee congruence.

(B) No, because angles alone do not guarantee congruence.

(C) Yes, because all three angles match.

(D) No, because the angles are in different order.

20. Point $(-3, -4)$ is reflected over the x-axis. What is the image?

21. If two figures are congruent, are they also similar?

(A) Yes, with scale factor $k = 1$.

(B) No, congruent and similar are different properties.

(C) Only if they are triangles.

(D) Only if they have the same orientation.

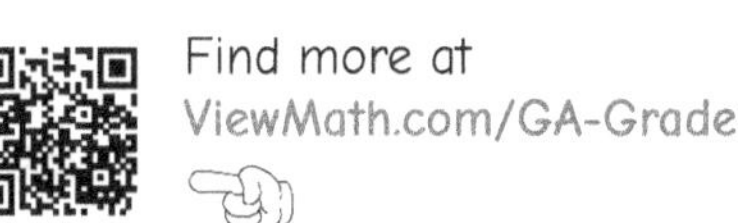

Find more at
ViewMath.com/GA-Grade8

22. Two parallel lines are cut by a transversal. One corresponding angle is 56°. What is the other?

Your Answer

23. A right triangle has legs 1 and 1. What is the exact length of the hypotenuse?

(A) 2

(B) $\sqrt{2}$

(C) 1

(D) $\sqrt{3}$

24. What is the distance between $(0,0)$ and $(5,0)$?

(A) 0

(B) 5

(C) 25

(D) $\sqrt{5}$

25. A basketball has a diameter of 24 cm. What is its volume? Use $\pi \approx 3.14$

(A) 7,234.6 cm^3

(B) 2,143.6 cm^3

(C) 904.3 cm^3

(D) 18,086.4 cm^3

26. Data: $(1,2), (2,4), (3,5), (4,8), (5,9)$. The dots appear to go upward from left to right. The association is:

(A) negative

(B) positive

(C) no association

(D) nonlinear

27. Data: $(1,3), (2,5), (3,4), (4,6), (5,8)$. A trend line through $(1,3)$ and $(5,8)$ has slope:

(A) $\frac{5}{4}$

(B) $\frac{4}{5}$

(C) 1

(D) 2

Find more at
ViewMath.com/GA-Grade8

28. *What is extrapolation?*

(A) *Making predictions within the range of the data*

(B) *Making predictions far beyond the range of the data*

(C) *Removing outliers from the data*

(D) *Finding the exact equation of the line*

29. *A survey asks students about their favorite sport (soccer or basketball) and grade (7th or 8th). 20 7th-graders chose soccer. This 20 is a:*

(A) *marginal frequency*

(B) *joint frequency*

(C) *relative frequency*

(D) *total frequency*

30. *What does the Mean Absolute Deviation (MAD) measure?*

(A) *The center of a data set*

(B) *The average distance of data points from the mean*

(C) *The median of the data*

(D) *The range of the data*

 # End of Practice Test 6 

Great job finishing the test!

 **My Score**

I got _____________ out of 30 questions right.

*Check your answers in the **Answer Key** at the back of the book.*

Review any questions you missed. That's how we learn!

Check Your Score Online!

*Visit **ViewMath Academy** to enter your answers and see which topics you need to review. You can also explore lessons, take quizzes, track your scores, and save your progress!*

viewmath.com/score/8.1.GA.21

Or go to viewmath.com/score and enter code: 8.1.GA.21

Practice Test 7

30 Questions

✏️ Before You Start ✏️

- ✓ **Read each question carefully** before choosing your answer.
- ✓ **Show your work** on scratch paper when you need to.
- ✓ **Skip hard questions** and come back to them later.
- ✓ **Check your answers** when you're done.
- ✓ **Take your time** — there's no rush!

⭐ You've Got This! ⭐

Do your best and show what you know!

1. Give an example of an irrational number between 1 and 2.

   Your Answer:

2. A student converts $0.\overline{36}$ and gets $\frac{36}{100}$. What mistake did the student make?

   (A) The student divided by 100 instead of 99.

   (B) The student forgot to simplify.

   (C) The student treated a repeating decimal as a terminating decimal.

   (D) The student multiplied by 10 instead of 100.

3. Estimate $\sqrt{3} + \sqrt{7}$ to one decimal place.

   Your Answer:

4. Which expression is equivalent to $\frac{9^3 \cdot 9^2}{9^4}$?

   (A) 9^{-1}

   (B) 9^0

   (C) 9^1

   (D) 9^{24}

5. A cube-shaped box has a volume of 216 cubic centimeters. What is the edge length of the box?

   Your Answer:

6. Which of the following is 350,000 written in scientific notation?

   (A) 35×10^4

   (B) 3.5×10^5

   (C) 3.5×10^4

   (D) 0.35×10^6

Find more at
ViewMath.com/GA-Grade8

ViewMath.com

7. Earth's mass is about 6×10^{24} kg and Jupiter's mass is about 1.9×10^{27} kg. About how many times more massive is Jupiter than Earth? Round to the nearest whole number.

Your Answer:

8. Which table does NOT represent a proportional relationship?

(A)
x	1	2
y	4	8

(B)
x	2	4
y	5	10

(C)
x	3	6
y	9	15

(D)
x	5	10
y	15	30

9. What is the slope of the line through $(1, 3)$ and $(4, 12)$?

(A) 2

(B) 3

(C) 4

(D) 9

10. Solve: $y = -2x + 10$ and $y = x + 1$.

(A) $(2, 3)$

(B) $(3, 4)$

(C) $(4, 5)$

(D) $(5, 0)$

11. A rectangle has a perimeter of 48 m. Its length is twice its width. Find the dimensions.

Your Answer:

12. In the ordered pair (x, y), x is called the:

   (A) output

   (B) range

   (C) input

   (D) function

13. A function rule is $f(x) = \frac{x}{2} + 6$. What is $f(8)$?

   (A) 7

   (B) 10

   (C) 14

   (D) 20

14. Two functions have the equations $y = 5x + 3$ and $y = 5x - 7$. Which statement is true?

   (A) They have the same rate of change and the same initial value.

   (B) They have different rates of change and different initial values.

   (C) They have the same rate of change but different initial values.

   (D) They have different rates of change but the same initial value.

15. Is $y = 8x + 1$ linear or nonlinear?

   Your Answer:

16. A linear function passes through $(0, 5)$ and $(2, 11)$. What is its equation?

   (A) $y = 2x + 5$

   (B) $y = 3x + 5$

   (C) $y = 5x + 3$

   (D) $y = 6x + 5$

17. A graph of distance vs. time goes upward from left to right. What does this tell you?

   (A) The object is slowing down.

   (B) The distance is decreasing.

   (C) The distance is increasing over time.

   (D) The object is standing still.

Find more at
ViewMath.com/GA-Grade8

ViewMath.com

18. A rectangle has vertices at $(1, 2)$, $(5, 2)$, $(5, 4)$, and $(1, 4)$. After a translation of 3 units right and 1 unit up, what are the coordinates of the vertex that was at $(1, 2)$?

(A) $(4, 3)$

(B) $(4, 1)$

(C) $(2, 5)$

(D) $(-2, 3)$

19. Square A has side length 4 cm. Square B has side length 6 cm. Are they congruent?

(A) Yes, because they are both squares.

(B) Yes, because all squares have $90°$ angles.

(C) No, because their side lengths differ.

(D) No, because their angles differ.

20. Dilate the point $(-4, 10)$ by a factor of $\frac{1}{2}$ from the origin. What is the image?

21. Are all squares similar to each other?

(A) No, because they can have different side lengths.

(B) No, because they can have different angles.

(C) Yes, because all squares have equal angles and proportional sides.

(D) Yes, but only if they have the same perimeter.

22. A right triangle has one angle of $32°$. What is the third angle?

(A) $148°$

(B) $58°$

(C) $68°$

(D) $48°$

23. A right triangle has legs 20 and 21. Find the hypotenuse.

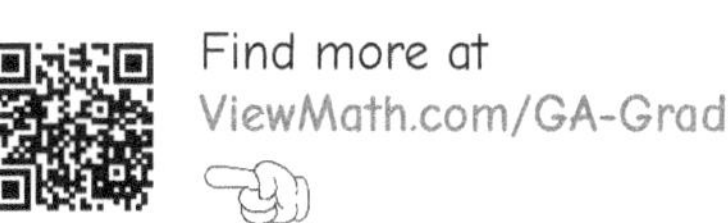

Find more at
ViewMath.com/GA-Grade8

24. Points P and Q are shown on the grid. What is the distance between them?

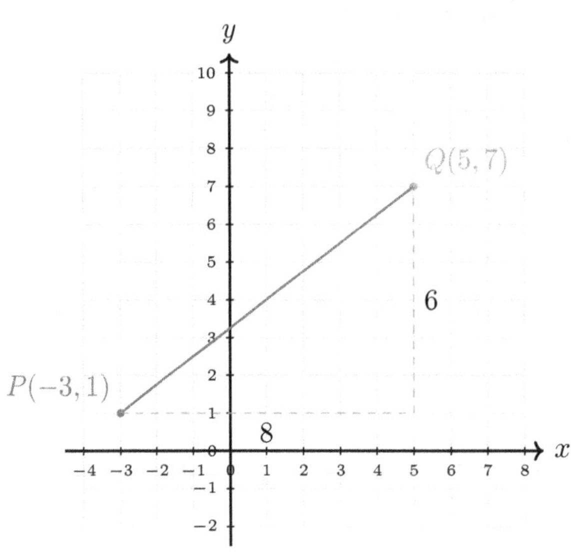

- (A) 14
- (B) $\sqrt{14}$
- (C) 10
- (D) $\sqrt{28}$

25. A sphere has volume $\frac{256}{3}\pi$ ft^3. What is the radius?

Your Answer

26. A scatter plot has a strong upward trend. Which r-value (correlation) is most likely?

- (A) $r = -0.9$
- (B) $r = 0.1$
- (C) $r = 0.9$
- (D) $r = 0$

27. *Data:* $(1, 8), (2, 7), (3, 5), (4, 4), (5, 2)$. *Draw a trend line and write its equation.*

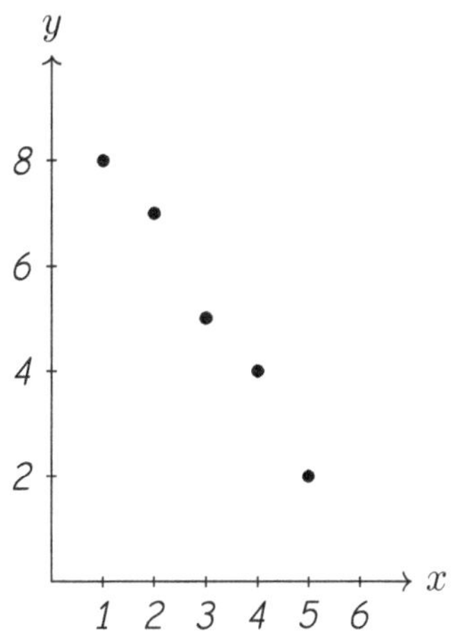

Your Answer:

28. *A linear model is* $y = 2x + 5$. *What does the slope represent?*

(A) The starting value

(B) The rate of change per unit increase in x

(C) The x-intercept

(D) The total value of y

29. *Using the table above, what proportion of pizza lovers are boys?*

(A) $\frac{18}{42}$

(B) $\frac{24}{42}$

(C) $\frac{24}{80}$

(D) $\frac{24}{40}$

30. *Data:* $\{8, 12, 10, 14, 6\}$. *Mean* $= 10$. *What is the MAD?*

(A) 2

(B) 2.4

(C) 3

(D) 10

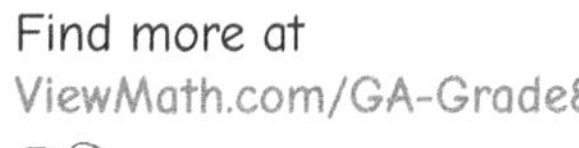
Find more at
ViewMath.com/GA-Grade8

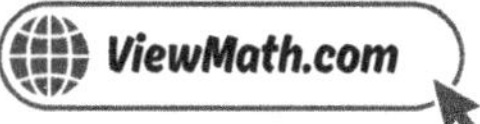

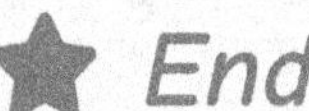 End of Practice Test 7 

Great job finishing the test!

 My Score

I got _______ out of 30 questions right.

*Check your answers in the **Answer Key** at the back of the book.*

Review any questions you missed. That's how we learn!

Check Your Score Online!

*Visit **ViewMath Academy** to enter your answers and see which topics you need to review. You can also explore lessons, take quizzes, track your scores, and save your progress!*

viewmath.com/score/8.1.GA.22

*Or go to **viewmath.com/score** and enter code: 8.1.GA.22*

Practice Test 8

☑ *30 Questions*

✏ Before You Start ✏

- ✓ **Read each question carefully** before choosing your answer.
- ✓ **Show your work** on scratch paper when you need to.
- ✓ **Skip hard questions** and come back to them later.
- ✓ **Check your answers** when you're done.
- ✓ **Take your time** — there's no rush!

★ *You've Got This!* ★

Do your best and show what you know!

1. Which of the following is a true statement?

   A) Every square root is irrational.

   B) Every integer is irrational.

   C) Every integer is rational.

   D) Every decimal is irrational.

2. The table below shows repeating decimals and their fraction equivalents. Find the missing fraction for $0.\overline{81}$.

Decimal	Unsimplified Fraction	Simplest Form
$0.\overline{27}$	$\frac{27}{99}$	$\frac{3}{11}$
$0.\overline{54}$	$\frac{54}{99}$	$\frac{6}{11}$
$0.\overline{81}$	?	?

Your Answer

3. A square patio has an area of 75 square feet. Estimate the side length to one decimal place.

Your Answer

4. Which shows $\frac{1}{2^4}$ written with a negative exponent?

   A) $(-2)^4$

   B) 2^{-4}

   C) -2^4

   D) 4^{-2}

5. Evaluate $\sqrt[3]{-125}$.

Your Answer

Get Online

Find more at
ViewMath.com/GA-Grade8

ViewMath.com

6. The bar chart compares the masses of four animals. Which animal's mass is closest to 1×10^2 kilograms?

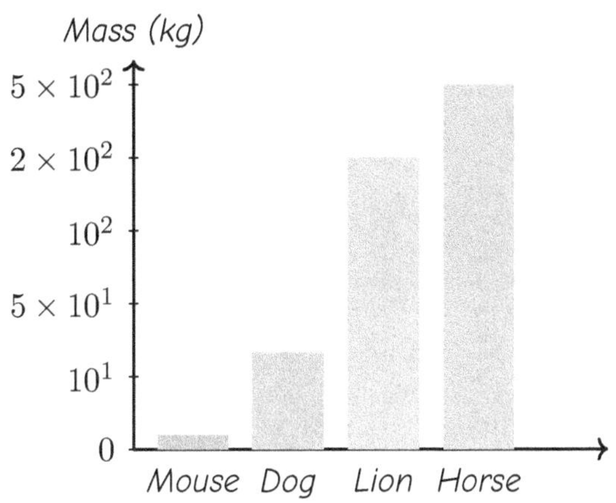

(A) Mouse

(B) Dog

(C) Lion

(D) Horse

7. A phone stores 6.4×10^{10} bytes. A photo uses 4×10^6 bytes. How many photos can the phone store?

(A) 1.6×10^3

(B) 1.6×10^4

(C) 1.6×10^{16}

(D) 2.4×10^4

8. Machine A fills $y = 12x$ bottles per hour. Machine B fills 50 bottles in 5 hours. Which machine is faster?

(A) Machine A

(B) Machine B

(C) They fill at the same rate.

(D) Cannot be determined.

9. The temperature drops from 68°F to 50°F over 6 hours. What is the rate of change in temperature?

(A) 3°F per hour

(B) −3°F per hour

(C) 6°F per hour

(D) −6°F per hour

10. Solve: $x + 2y = 14$ and $x + y = 9$.

Your Answer:

11. The table shows total costs for two cell phone companies.

Months	Company X	Company Y
1	$35	$50
2	$60	$65
3	$85	$80
4	$110	$95

Between which two months do the costs become equal?

(A) Between month 1 and month 2

(B) Between month 2 and month 3

(C) Between month 3 and month 4

(D) They are never equal.

12. Which of these real-world examples is a function?

(A) A person's age mapped to their height (a person can be the same height at different ages)

(B) A student's name mapped to their student ID number

(C) A city mapped to all the people who live there

(D) A month mapped to every holiday in that month

13. If $h(x) = -3x + 15$, for what value of x is $h(x) = 0$?

Your Answer:

Find more at
ViewMath.com/GA-Grade8

14. The tables below show two functions. Find the rate of change and initial value of each. Which function will have a greater value at $x = 10$?

Function A

x	y
0	20
1	23
2	26
3	29

Function B

x	y
0	5
1	10
2	15
3	20

Your Answer

15. A table shows $(1, 1)$, $(2, 8)$, $(3, 27)$, $(4, 64)$. Is this linear or nonlinear? What pattern do you see?

Your Answer

16. A function has a slope of -3 and passes through $(2, 1)$. What is the y-intercept?

Your Answer

17. A graph is flat for 4 seconds on a speed-time graph. What does this mean?

Your Answer

18. A point at $(-4, 3)$ is rotated $180°$ around the origin. What are the new coordinates?

(A) $(4, 3)$ (B) $(-4, -3)$

(C) $(4, -3)$ (D) $(3, -4)$

19. Describe a sequence of rigid transformations that maps a triangle with vertices $(1, 1)$, $(4, 1)$, $(1, 3)$ to a triangle with vertices $(-1, 1)$, $(-4, 1)$, $(-1, 3)$.

Your Answer:

20. Square $ABCD$ is shown. It is dilated by a factor of 2 from the origin to form $A'B'C'D'$. What are the coordinates of C'?

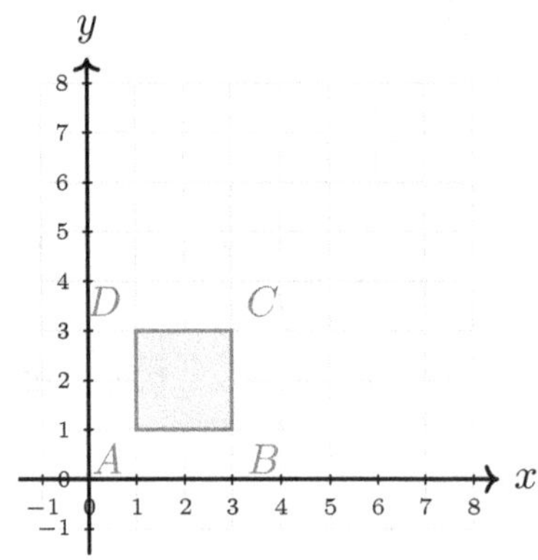

A) $(5, 5)$

B) $(6, 3)$

C) $(6, 6)$

D) $(3, 6)$

21. A square has side 5. After a dilation by $k = 4$, what is the perimeter of the image?

Your Answer:

22. A triangle has angles $48°$ and $67°$. What is the third angle?

A) $65°$

B) $75°$

C) $55°$

D) $115°$

Find more at
ViewMath.com/GA-Grade8

ViewMath.com

23. A right triangle has hypotenuse 13 and one leg 5. What is the other leg?

(A) 8

(B) 18

(C) 12

(D) $\sqrt{8}$

24. Find the distance between $(0,0)$ and $(2,3)$. Leave your answer in simplest radical form.

Your Answer:

25. A cone has volume 100 cm^3. A cylinder with the same base and height has volume ___ cm^3.

Your Answer:

26. In a scatter plot, several data points bunch together near $(3, 20)$. This grouping is called:

(A) an outlier

(B) a gap

(C) a cluster

(D) a negative association

27. Data: $(1, 2), (2, 4), (3, 7), (4, 8), (5, 11)$. Estimate the slope of a reasonable trend line.

Your Answer:

28. In the model $y = 6x + 12$, if x increases by 3 units, how much does y increase?

(A) 3

(B) 6

(C) 18

(D) 30

Find more at
ViewMath.com/GA-Grade8

29. Using the table from q19, what percentage of boys prefer reading? What percentage of girls prefer reading? Is there an association?

Your Answer

30. A MAD of 0 tells you that:

(A) The data values are all different

(B) All data values are the same

(C) The mean is 0

(D) There is one data point

 # End of Practice Test 8 

Great job finishing the test!

 My Score

I got _____________ out of 30 questions right.

Check your answers in the **Answer Key** at the back of the book.

💡 Review any questions you missed. That's how we learn!

📊 Check Your Score Online!

Visit **ViewMath Academy** to enter your answers and see which topics you need to review. You can also explore lessons, take quizzes, track your scores, and save your progress!

viewmath.com/score/8.1.GA.23

Or go to viewmath.com/score and enter code: 8.1.GA.23

9

Practice Test 9

 *30 Questions*

✏️ Before You Start ✏️

- ✔ **Read each question carefully** before choosing your answer.
- ✔ **Show your work** on scratch paper when you need to.
- ✔ **Skip hard questions** and come back to them later.
- ✔ **Check your answers** when you're done.
- ✔ **Take your time** — there's no rush!

 You've Got This! 

Do your best and show what you know!

1. Which of the following could NOT be the decimal expansion of a rational number?

(A) 2.500

(B) $0.\overline{9}$

(C) 1.41421356… (non-repeating)

(D) $0.\overline{36}$

2. What is $0.\overline{4}$ written as a fraction?

(A) $\frac{4}{10}$

(B) $\frac{4}{9}$

(C) $\frac{2}{5}$

(D) $\frac{1}{4}$

3. Between which two consecutive integers does $3\sqrt{2}$ lie?

(A) 3 and 4

(B) 4 and 5

(C) 5 and 6

(D) 6 and 7

4. Which expression equals $\frac{1}{9}$?

(A) 3^{-2}

(B) 3^{-3}

(C) 9^{-2}

(D) $(-3)^2$

5. Between which two consecutive whole numbers does $\sqrt{50}$ lie?

(A) 6 and 7

(B) 7 and 8

(C) 24 and 26

(D) 8 and 9

6. Which is larger: 8×10^5 or 3×10^6?

(A) 8×10^5, because $8 > 3$

(B) 3×10^6, because the exponent is larger

(C) They are equal

(D) It cannot be determined

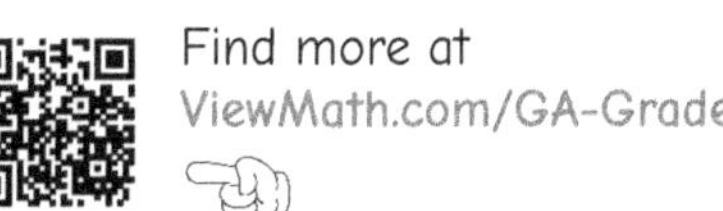

7. *Simplify $(7 \times 10^{-3})(8 \times 10^{-2})$. Write in proper scientific notation.*

   (A) 56×10^{-5} (B) 5.6×10^{-5}

   (C) 5.6×10^{-4} (D) 5.6×10^{6}

8. Two delivery services are compared. Service P delivers $y = 4x$ packages per hour. Service Q's data is shown in the table.

Hours (x)	2	4	6	8
Packages (y)	10	20	30	40

How many more packages does Service Q deliver than Service P in 10 hours?

Your Answer:

9. Find the slope through $(0, -2)$ and $(3, 7)$.

   (A) 3   (B) -3

   (C) $\frac{5}{3}$ (D) $\frac{9}{3}$

10. Use the tables to find the solution to the system of equations.

Equation 1: $y = 2x + 1$

x	y
0	1
1	3
2	5
3	7
4	9

Equation 2: $y = -x + 7$

x	y
0	7
1	6
2	5
3	4
4	3

Your Answer:

11. A farmer has chickens and cows. There are 20 heads and 56 legs. How many cows are there?

(A) 6

(B) 8

(C) 10

(D) 12

12. Which set of ordered pairs represents a function?

(A) $\{(1,2),(1,3),(2,4)\}$

(B) $\{(3,5),(4,5),(5,5)\}$

(C) $\{(2,6),(2,8),(3,6)\}$

(D) $\{(0,1),(0,-1),(1,2)\}$

13. A student says $f(3) = f \times 3$. What is the student's mistake?

(A) $f(3)$ means "f divided by 3."

(B) $f(3)$ means "the function f evaluated at 3," not multiplication.

(C) $f(3)$ means "3 is the function."

(D) There is no mistake.

14. Function A: $y = 3x + 8$. Function B: $y = 3x + 2$. Which function has a greater initial value?

(A) Function A

(B) Function B

(C) They have the same initial value.

(D) Cannot be determined.

15. The graph of a function is a straight line with slope -3. This function is:

(A) Nonlinear, because the slope is negative.

(B) Linear, because the graph is a straight line.

(C) Nonlinear, because it is decreasing.

(D) Cannot be determined.

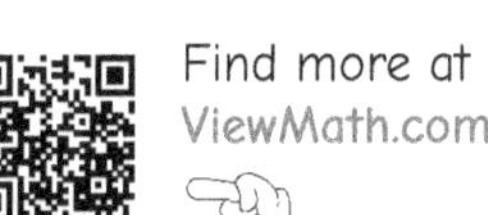

Find more at
ViewMath.com/GA-Grade8

16. The table below shows a linear function. What is the equation?

x	0	1	2	3
y	-4	-1	2	5

(A) $y = 3x - 4$

(B) $y = -4x + 3$

(C) $y = -3x + 4$

(D) $y = 3x + 4$

17. A ball is thrown straight up. Its height over time:

(A) Increases, then decreases — the graph is non-linear.

(B) Increases, then decreases — the graph is linear.

(C) Decreases only — the graph is linear.

(D) Increases only — the graph is nonlinear.

18. Which of the following is NOT preserved under a translation?

(A) Side lengths

(B) Angle measures

(C) Position

(D) Parallelism of sides

19. Which transformation would NOT preserve congruence?

(A) Reflection over the x-axis

(B) Translation 2 units up

(C) Dilation by a factor of 1.5

(D) Rotation of 270°

20. Which transformation is equivalent to reflecting over the x-axis and then reflecting over the y-axis?

(A) Translation

(B) 90° rotation

(C) 180° rotation

(D) Dilation by -1

Find more at
ViewMath.com/GA-Grade8

21. A dilation centered at the origin maps $(4, -2)$ to $(12, -6)$. What is the scale factor?

(A) 2

(B) 3

(C) 4

(D) 6

22. What is the sum of the interior angles of any triangle?

(A) 90°

(B) 180°

(C) 270°

(D) 360°

23. Which set of side lengths forms a right triangle?

(A) 3, 4, 6

(B) 5, 12, 13

(C) 4, 5, 7

(D) 2, 3, 4

24. The distance formula is derived from:

(A) The area formula for a rectangle

(B) The Pythagorean Theorem

(C) The perimeter formula

(D) The midpoint formula

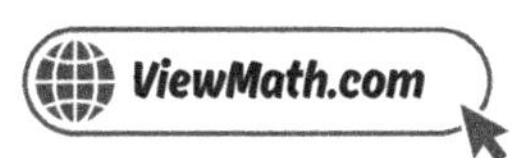

25. What is the volume of the cylinder shown below? Use $\pi \approx 3.14$.

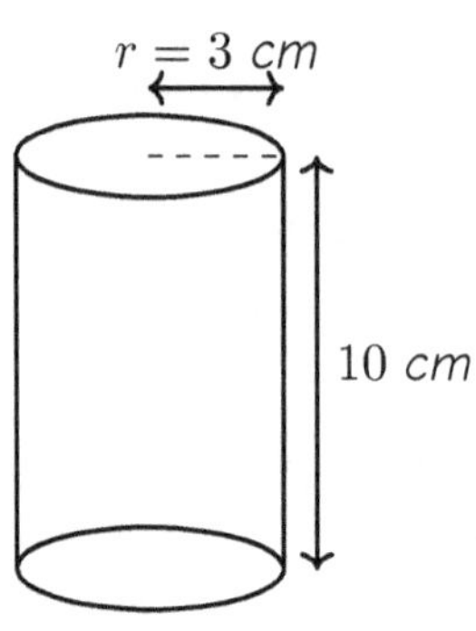

(A) $94.2\ cm^3$

(B) $188.4\ cm^3$

(C) $282.6\ cm^3$

(D) $565.2\ cm^3$

26. True or false: A scatter plot can show both clustering and outliers at the same time.

(A) True — they describe different features of the data.

(B) False — data has either clusters or outliers, not both.

(C) True — but only if there is no association.

(D) False — outliers are always in clusters.

27. A trend line is drawn through the scatter plot. What is the approximate slope of the line?

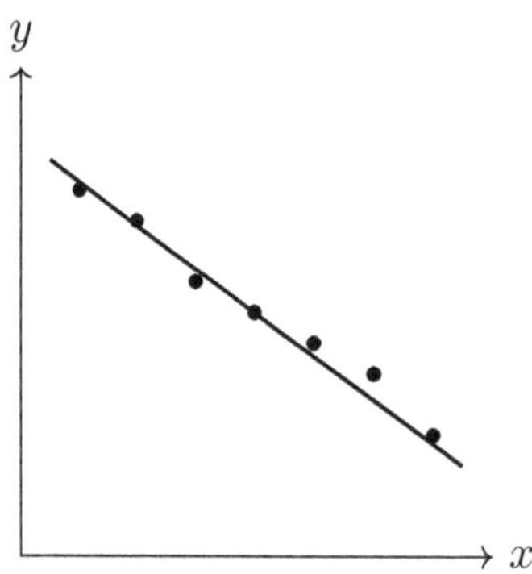

(A) $\approx -\frac{5}{7}$

(B) $\approx \frac{5}{7}$

(C) ≈ -2

(D) ≈ 1

28. A model gives $y = 5x + 20$ where x is weeks and y is savings (dollars). If $x = 8$, what is the predicted savings?

(A) $45

(B) $60

(C) $68

(D) $40

29. Which of these is NOT a question you can answer with a two-way table?

(A) What fraction of boys prefer soccer?

(B) Is there an association between grade and preferred lunch?

(C) What is the mean score on the math test?

(D) How many 8th graders chose art?

30. Data: $\{100, 105, 95, 110, 90\}$. Calculate the MAD.

Your Answer

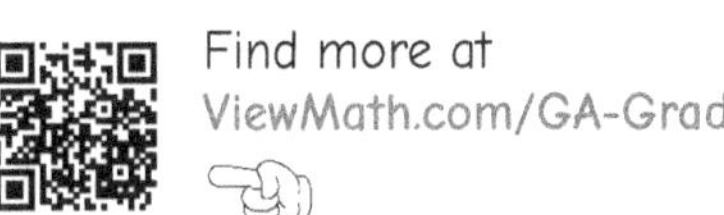

Find more at
ViewMath.com/GA-Grade8

 # End of Practice Test 9 

Great job finishing the test!

 **My Score**

I got _____________ out of 30 questions right.

*Check your answers in the **Answer Key** at the back of the book.*

Review any questions you missed. That's how we learn!

Check Your Score Online!

Visit **ViewMath Academy** to enter your answers and see which topics you need to review. You can also explore lessons, take quizzes, track your scores, and save your progress!

viewmath.com/score/8.1.GA.24

Or go to viewmath.com/score and enter code: 8.1.GA.24

Practice Test 10

☑ 30 Questions

✏ Before You Start ✏

- ✓ **Read each question carefully** before choosing your answer.

- ✓ **Show your work** on scratch paper when you need to.

- ✓ **Skip hard questions** and come back to them later.

- ✓ **Check your answers** when you're done.

- ✓ **Take your time** — there's no rush!

⭐ You've Got This! ⭐

Do your best and show what you know!

1. *Look at the Venn diagram below. In which region does $\sqrt{5}$ belong?*

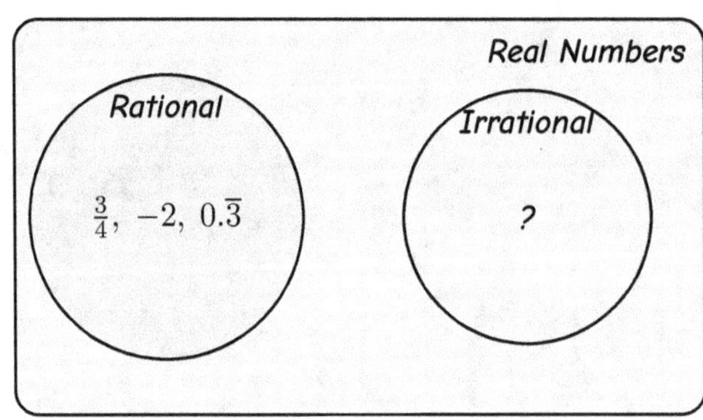

  (A) *Rational region, because $\sqrt{5} = 2.5$*

  (B) *Irrational region, because 5 is not a perfect square*

  (C) *Rational region, because $\sqrt{5}$ is a square root*

  (D) *Neither region, because $\sqrt{5}$ is not a real number*

2. *The steps below show a student's work for converting $0.\overline{2}$ to a fraction. Which step contains the error?*

Step	Work
1	Let $x = 0.222\ldots$
2	$10x = 2.222\ldots$
3	$10x - x = 2.222\ldots - 0.222\ldots$
4	$9x = 2$
5	$x = \frac{2}{10}$

  (A) *Step 2*

  (B) *Step 3*

  (C) *Step 4*

  (D) *Step 5*

3. *Which is the best estimate for π^2?*

  (A) 6.3

  (B) 9.9

  (C) 3.1

  (D) 12.6

4. Simplify $\left(\frac{3}{4}\right)^{-2}$.

(A) $\frac{9}{16}$

(B) $\frac{-9}{16}$

(C) $\frac{16}{9}$

(D) $\frac{-6}{8}$

5. Which statement about $\sqrt{2}$ is true?

(A) $\sqrt{2}$ is a rational number

(B) $\sqrt{2}$ is an integer

(C) $\sqrt{2}$ is an irrational number

(D) $\sqrt{2} = 1.5$

6. The diagram below shows three cells viewed under a microscope, with their actual sizes labeled. Write all three sizes in scientific notation, then list them from smallest to largest.

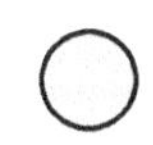
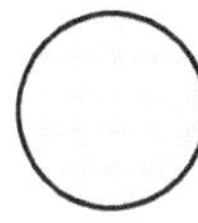
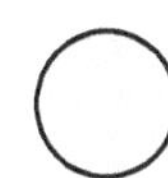

Cell A
0.008 mm

Cell B
0.05 mm

Cell C
0.0003 mm

Your Answer

7. Compute $\frac{3.6 \times 10^{10}}{1.2 \times 10^{4}}$. Write your answer in scientific notation.

Your Answer

8. If $y = kx$ and $y = 54$ when $x = 6$, find y when $x = 11$.

Your Answer

9. *A line goes down from left to right. Its slope must be:*

  (A) *Positive*   (B) *Negative*

  (C) *Zero*   (D) *Undefined*

10. *Solve using elimination:* $x + y = 10$ *and* $x - y = 4$

  (A) $(5, 5)$ (B) $(7, 3)$

  (C) $(6, 4)$ (D) $(8, 2)$

11. *A store sells T-shirts for \$12 and hats for \$8. Use the information in the table to find how many of each were sold.*

	Value
Total items sold	15
Total revenue	\$148

Your Answer

12. *A circle drawn on a coordinate plane:*

  (A) *is always a function*   (B) *is a function only if it is small*

  (C) *is never a function*   (D) *is a function if it is centered at the origin*

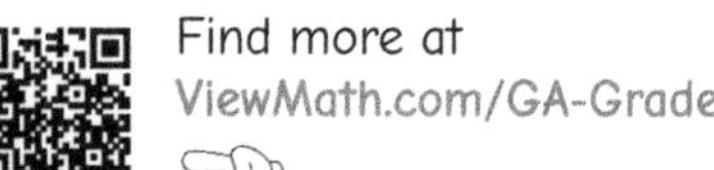

Find more at
ViewMath.com/GA-Grade8

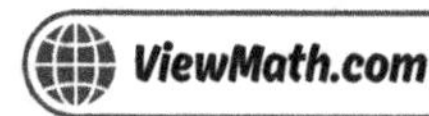

13. *Use the graph of f below to find $f(1) + f(4)$.*

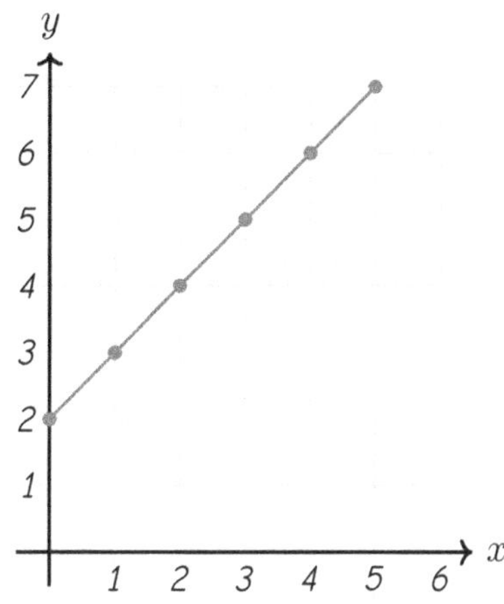

Your Answer:

14. *Function G is shown in the table:*

x	0	1	2	3	4
y	2	5	8	11	14

Function H: $y = 4x + 1$. Which function has the greater value at $x = 4$?

(A) Function G

(B) Function H

(C) They are equal.

(D) Cannot be determined.

15. *What is the form of a linear function?*

(A) $y = ax^2 + bx + c$

(B) $y = mx + b$

(C) $y = a^x$

(D) $y = \frac{a}{x}$

16. *Write the equation of a line with slope 6 and y-intercept −3.*

Your Answer:

17. *Which graph could represent a person standing still for 5 seconds, then walking at a steady pace?*

(A) A line sloping upward for the entire time

(B) A horizontal line, then a line sloping upward

(C) A line sloping downward, then a horizontal line

(D) A curve going upward the entire time

18. *Point A is shown on the coordinate plane below. Find the coordinates of A' after a 90° counterclockwise rotation around the origin.*

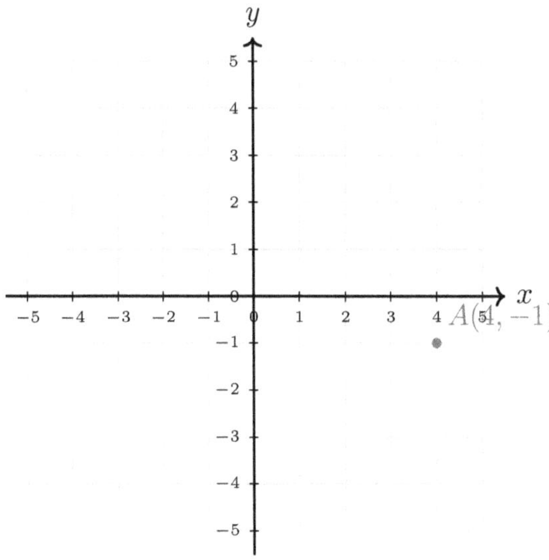

Your Answer:

19. *Triangle $ABC \cong$ Triangle DEF. If $AB = 8$ cm, what is DE?*

(A) 4 cm

(B) 8 cm

(C) 16 cm

(D) Cannot be determined

Find more at
ViewMath.com/GA-Grade8

20. *What is the image of $(2, -7)$ after a $180°$ rotation around the origin?*

(A) $(-2, 7)$ (B) $(2, 7)$

(C) $(-2, -7)$ (D) $(7, -2)$

21. *Two figures are similar if one can be mapped onto the other by a sequence of:*

(A) Only translations   (B) Only dilations

(C) Rigid transformations and dilations   (D) Only reflections

22. *In an isosceles triangle, the two base angles are each $54°$. What is the vertex angle?*

(A) $54°$ (B) $72°$

(C) $108°$ (D) $126°$

23. *A rectangular room is 12 ft long and 5 ft wide. What is the diagonal distance across the floor?*

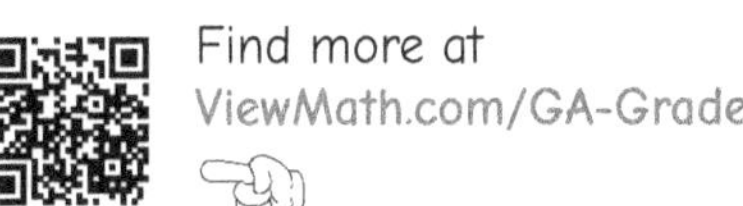

24. Which pair of points has a distance of exactly 13?

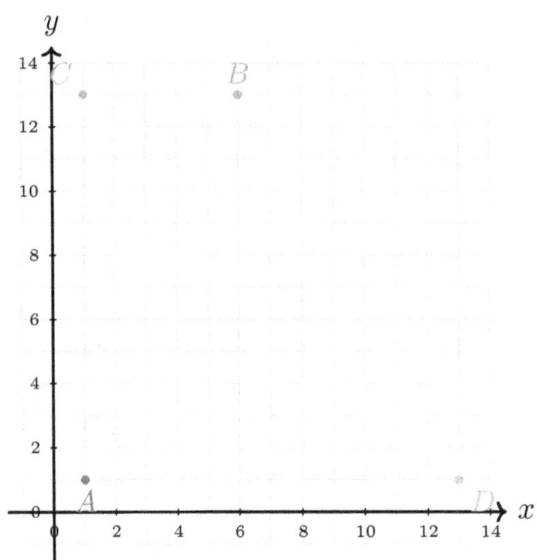

A. $A(1,1)$ and $B(6,13)$

B. $A(1,1)$ and $C(1,13)$

C. $A(1,1)$ and $D(13,1)$

D. $B(6,13)$ and $D(13,1)$

25. A cylinder has volume 200π cm^3 and height 8 cm. What is the radius?

A. 25 cm

B. 5 cm

C. $\sqrt{25}$ cm

D. 10 cm

26. A scatter plot of age (x) versus height (y) for people aged 0–80 would likely show what shape?

Your Answer

27. When drawing an informal line of best fit, about how many data points should be above the line?

A. All of them

B. None of them

C. About half

D. Exactly one

Find more at
ViewMath.com/GA-Grade8

28. *A scatter plot with trend line $y = 2x + 1$ is shown. What is the predicted value at $x = 4$?*

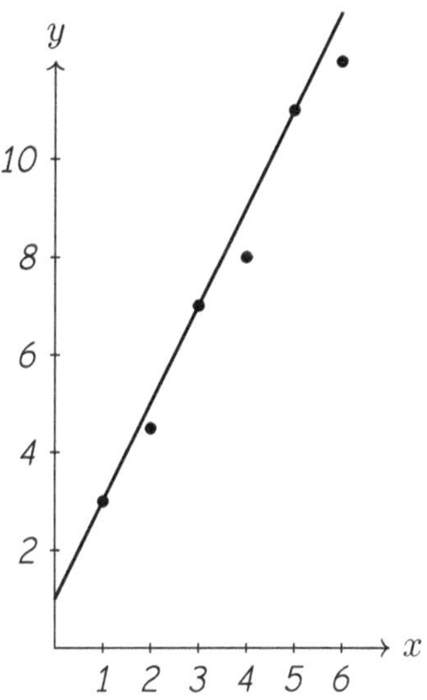

(A) 7

(B) 8

(C) 9

(D) 10

29.

	Pizza	Tacos	Total
Boys	24	16	40
Girls	18	22	40
Total	42	38	80

What is the relative frequency of boys who prefer pizza out of all 80 students?

(A) 0.30

(B) 0.60

(C) 0.20

(D) 0.525

30. *Set A: $\{3, 5, 7, 9, 11\}$. Set B: $\{6, 7, 7, 7, 8\}$. Without computing, which likely has a larger MAD?*

(A) Set A

(B) Set B

(C) Both are equal

(D) Cannot tell without computing

 # End of Practice Test 10 

Great job finishing the test!

 My Score

I got ___________ out of 30 questions right.

*Check your answers in the **Answer Key** at the back of the book.*

Review any questions you missed. That's how we learn!

Check Your Score Online!

*Visit **ViewMath Academy** to enter your answers and see which topics you need to review. You can also explore lessons, take quizzes, track your scores, and save your progress!*

viewmath.com/score/8.1.GA.25

Or go to viewmath.com/score and enter code: 8.1.GA.25

Answer Key & Explanations

Answer Key

First try each test on your own, then check your work here.

✅ Practice Test 1 — Answer Key

 False
 C
 B
 D
 B
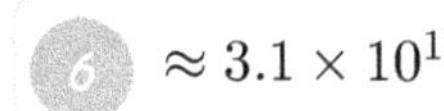 $\approx 3.1 \times 10^{12}$
 C
 B
 3

 $(9, 6)$
 B
 Yes, it is a function.
 5
 A
 $y = x^2$
 B

 Increasing, decreasing, increasing.

 $(0, -4)$
 C
 B
 D
 C
 B
 B
 A
 B
 B

28 At $x = 5$: $y = 30$. At $x = 30$: $y = -45$. The prediction at $x = 5$ is more reliable because it is within the data ran

29 C
30 Mean $= 4$. MAD $= 2.4$. The outlier (12) pulls the mean up and increases the MAD.

💡 Time to Learn! 💡

*Review the explanations below, **especially for the questions you missed.***

Understanding why each answer is correct builds stronger problem-solving skills.

Tip: *Circle any questions you got wrong, then read their explanation carefully.*

📖 Practice Test 1 — Detailed Explanations

Find more at
ViewMath.com/GA-Grade8

🌐 **ViewMath.com**

1. $\frac{22}{7} \approx 3.142857\ldots$ is a rational approximation of π, but $\pi = 3.14159265\ldots$ is irrational. They are close but not equal.

2. $0.\overline{123}$ has a 3-digit repeating block, so you should multiply by $10^3 = 1000$, not 100.

3. Area $= \sqrt{20} \times \sqrt{8} = \sqrt{160}$. $\sqrt{160} \approx 12.65$ cm^2. (Note: $\sqrt{160} = 4\sqrt{10} \approx 12.65$.) Choices A and C confuse addition with multiplication.

4. $7^5 \cdot 7^{-5} = 7^{5+(-5)} = 7^0 = 1$.

5. If $s = 8$, then $A = 8^2 = 64$. If $e = 4$, then $V = 4^3 = 64$. Since $A = V = 64$, the pair $s = 8, e = 4$ works.

6. $498{,}000{,}000 \approx 5 \times 10^8$ and $6.200 \approx 6.2 \times 10^3$. Product $\approx 5 \times 6.2 \times 10^{11} = 31 \times 10^{11} = 3.1 \times 10^{12}$.

7. $(3 \times 10^4)(5 \times 10^3) = 15 \times 10^7 = 1.5 \times 10^8$.

8. At \$8 per hour with no extra fee, the relationship is proportional: $y = 8x$.

9. $m = \frac{11-(-1)}{7-3} = \frac{12}{4} = 3$.

10. Add: $2x = 18$, $x = 9$. Then $y = 15 - 9 = 6$.

11. Worker A: $y = 10x + 20$, so at 6 hrs: \$80. Worker B: $y = 15x$, so at 6 hrs: \$90. Difference: \$10.

12. The graph is a straight line. Any vertical line will cross it at exactly one point, so it passes the vertical line test and is a function

13. $f(6) = \frac{6+4}{2} = \frac{10}{2} = 5$.

Find more at
ViewMath.com/GA-Grade8

14 The rate of change is the slope. Function A has slope 5 and Function B has slope 2. Since $5 > 2$, Function A has the greater rate of change.

15 Any equation where x has an exponent other than 1 (such as x^2, x^3, $\frac{1}{x}$, $\sqrt{x}$) is nonlinear.

16 The pool starts at 150 gallons ($b = 150$) and gains 10 gallons per minute ($m = 10$). So $y = 10x + 150$.

17 Filling: the water level rises. Drinking: the level drops. Refilling: the level rises again.

18 A $180°$ rotation uses $(x, y) \to (-x, -y)$. So $(0, 4) \to (0, -4)$.

19 By definition, congruent figures can be mapped onto each other using rigid transformations (translations, reflections, rotations).

20 Original sides: 6, 8, and $\sqrt{6^2 + 8^2} = 10$. Perimeter $= 6 + 8 + 10 = 24$. Dilation by $\frac{1}{2}$ multiplies every length by $\frac{1}{2}$, so the new perimeter is $24 \times \frac{1}{2} = 12$.

21 The longest side is 13. Multiplied by $k = 2$: $13 \times 2 = 26$.

22 Corresponding angles (same position at each intersection) are equal when the lines are parallel.

23 $7^2 + 24^2 = 49 + 576 = 625 = 25^2$. Since $a^2 + b^2 = c^2$, it is a right triangle.

24 The distance formula is $d = \sqrt{(x_2 - x_1)^2 + (y_2 - y_1)^2}$.

25 $V = \pi r^2 h = 3.14 \times 25 \times 12 = 942 \ ft^3$.

26 The dots go from upper-left to lower-right in a roughly straight pattern. This is a negative linear association.

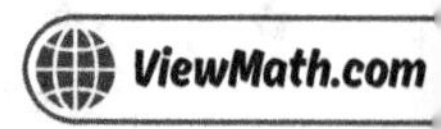

27 Outliers should generally be ignored when drawing the line of best fit, as they don't represent the overall trend.

28 $y = -3(5) + 45 = 30$ (interpolation). $y = -3(30) + 45 = -45$ (extrapolation — unreliable and gives a negative value that may not make sense).

29 $\frac{25}{40} = 0.625 = 62.5\%.$

30 Deviations: $2, 2, 2, 2, 8.$ MAD $= \frac{2+2+2+2+8}{5} = \frac{16}{5} = 3.2.$ Without 12, all values are 2, MAD $= 0.$

✅ Practice Test 2 — Answer Key

1 C **2** Multiply by 100; $100x = 54.\overline{54}$; $99x = 54$; $x = \frac{54}{99} = \frac{6}{11}$ **3** ≈ 0.02 **4** D **5** B

6 B **7** A **8** 180 mL **9** B **10** $(3, 1)$ **11** C **12** $\{(1,2),(2,3),(3,4)\}$ **13** 3

14 6 hours **15** B **16** A **17** C **18** B **19** 5 cm **20** B **21** C **22** 60°

23 C **24** 10 **25** B

26 Positive linear association. Outlier at approximately $(1, 7)$ — far above the trend. **27** C **28** C

29 B **30** A

💡 Time to Learn! 💡

Review the explanations below, **especially for the questions you missed**.

Understanding why each answer is correct builds stronger problem-solving skills.

Tip: Circle any questions you got wrong, then read their explanation carefully.

Find more at
ViewMath.com/GA-Grade8

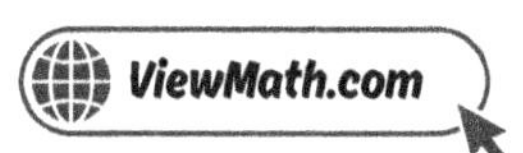

📖 Practice Test 2 — Detailed Explanations

1. $P = \sqrt{2}$ is irrational (since 2 is not a perfect square) and $Q = 3$ is an integer, which is rational.

2. The repeating block has 2 digits, so multiply by 100. Then $100x - x = 99x = 54$, giving $x = \frac{54}{99} = \frac{6}{11}$.

3. $\pi \approx 3.1416$ and $\sqrt{10} \approx 3.1623$. The distance is $3.1623 - 3.1416 = 0.0207 \approx 0.02$.

4. A negative exponent means reciprocal: $4^{-3} = \frac{1}{4^3} = \frac{1}{64}$.

5. $x = \pm\sqrt{\frac{9}{16}} = \pm\frac{3}{4}$, since $\left(\frac{3}{4}\right)^2 = \frac{9}{16}$.

6. $\frac{4.2 \times 10^3}{4.2 \times 10^{-3}} = 10^{3-(-3)} = 10^6$. So 4.2×10^3 is 10^6 (one million) times as large.

7. Divide the coefficients: $\frac{8}{2} = 4$. Subtract the exponents: $10^{7-3} = 10^4$. Answer: 4×10^4.

8. Rate $= \frac{45}{15} = 3$ mL/min. In 60 minutes: $3 \times 60 = 180$ mL.

9. Rate of change $= \frac{16-4}{6} = \frac{12}{6} = 2$ cm per week.

10. Substitute: $3x + (x - 2) = 10$, so $4x = 12$, $x = 3$. Then $y = 3 - 2 = 1$.

11. $a + o = 12$ and $1.50a + 1.00o = 14.50$. From first: $o = 12 - a$. Substitute: $1.50a + 12 - a = 14.50$, so $0.50a = 2.50$, $a = 5$.

12. Any set where each input appears only once is correct. For example, $\{(1,2),(2,3),(3,4)\}$ has no repeated inputs.

Find more at
ViewMath.com/GA-Grade8

13 Follow the dashed line at $y = 6$ until it meets the graph. That point is $(3, 6)$, so $x = 3$.

14 $8x + 20 = 3x + 50 \Rightarrow 5x = 30 \Rightarrow x = 6$.

15 $y = 7$ means y is always 7 regardless of x. This is a horizontal line, which is linear ($m = 0$, $b = 7$).

16 The slope $m = 50$ represents the amount added per week, so the weekly deposit is \$50.

17 Filling: water level rises (increasing). Soaking: level stays the same (constant). Draining: level drops (decreasing).

18 Translations preserve all side lengths. The longest side is still 8 cm.

19 $P = 2l + 2w$. So $28 = 2(9) + 2w$, giving $28 = 18 + 2w$, $2w = 10$, $w = 5$ cm.

20 The 90° CCW rule is $(x, y) \to (-y, x)$. So $(-3, 5) \to (-5, -3)$.

21 Multiply the side by the scale factor. $3 \times 4 = 12$ cm.

22 $\frac{180}{3} = 60°$. Each angle of an equilateral triangle is 60°.

23 $c^2 = 9^2 + 12^2 = 81 + 144 = 225$, so $c = 15$.

24 $d = \sqrt{(5 - (-1))^2 + (7 - (-1))^2} = \sqrt{36 + 64} = \sqrt{100} = 10$.

25 $V = \frac{1}{3}\pi(4^2)(9) = \frac{1}{3}\pi(144) = 48\pi$ cm^3.

Find more at
ViewMath.com/GA-Grade8

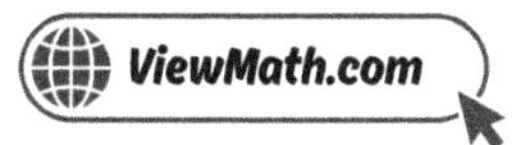

26. Direction: positive (up-right). Shape: linear. The point $(1, 7)$ is an outlier (most students who studied 1 hour scored around 3, not 7).

27. A line of best fit (trend line) is a straight line drawn to approximate the overall trend, coming close to most points.

28. A slope of 0 means no change in y as x changes; the line is horizontal.

29. 60% of boys chose band vs. 45% of girls. The difference in conditional proportions shows an association.

30. Deviations: $|3 - 6| = 3, |5 - 6| = 1, |5 - 6| = 1, |7 - 6| = 1, |10 - 6| = 4.$ Sum $= 10.$ MAD should be checked: mean $= (3 + 5 + 5 + 7 + 10)/5 = 6.$ MAD $= 10/5 = 2.$ Actually 2.

✅ Practice Test 3 — Answer Key

 C    C    C    B    C    A    A    A    B    D

 B    No    C    Q    A

 B    B    C    C    C    D    $x = 25$  C    C    C

26. Yes. Example: population growth over time may curve upward (exponential), showing a positive but nonlinear

 B    B   29 B   30 B

💡 Time to Learn! 💡

*Review the explanations below, **especially for the questions you missed**.*

Understanding why each answer is correct builds stronger problem-solving skills.

Tip: *Circle any questions you got wrong, then read their explanation carefully.*

📖 Practice Test 3 — Detailed Explanations

1. $0.\overline{81}$ is a repeating decimal, so it can be written as a fraction: $0.\overline{81} = \frac{81}{99} = \frac{9}{11}$. The others are irrational.

2. Let $x = 0.\overline{27}$. Then $100x = 27.\overline{27}$. Subtract: $99x = 27$, so $x = \frac{27}{99} = \frac{3}{11}$.

3. $4\sqrt{2} \approx 4 \times 1.414 = 5.656$ and $\sqrt{30} \approx 5.477$. Since $5.656 > 5.477$, $4\sqrt{2}$ is larger.

4. $\frac{8^6}{8^6} = 8^{6-6} = 8^0 = 1$. The expression equals 8^0.

5. $\text{Side} = \sqrt{196} = 14$ cm, since $14 \times 14 = 196$.

6. Move the decimal 4 places right: $0.00072 = 7.2 \times 10^{-4}$.

7. $(9.5 \times 10^{-18})(2 \times 10^6) = 19 \times 10^{-12} = 1.9 \times 10^{-11}$ g.

8. Store A: $k = 2$ per pound. Store B: $k = \frac{7.50}{3} = 2.50$ per pound. Store A is cheaper.

9. $\frac{270-120}{5-2} = \frac{150}{3} = 50$ mph.

Find more at
ViewMath.com/GA-Grade8

10 Rewrite the second: $-2y = -4x - 10$, so $y = 2x + 5$. Both equations are the same line. Infinitely many solutions.

11 $h + b = 80$ and $3h + 5b = 340$. From first: $b = 80 - h$. Substitute: $3h + 5(80 - h) = 340$, so $-2h + 400 = 340$, $-2h = -60$, $h = 30$.

12 Input a maps to two different outputs (5 and 12), so the relation cannot be a function.

13 Set $4x - 3 = 17$. Add 3: $4x = 20$. Divide by 4: $x = 5$.

14 Function Q has slope 7, which is greater than Function P's slope of 3.

15 $10 - 5 = 5$, $15 - 10 = 5$, $20 - 15 = 5$. The constant rate of change of 5 confirms the function is linear.

16 Slope: $\frac{20-8}{3-0} = \frac{12}{3} = 4$. y-intercept: 8. So $y = 4x + 8$.

17 $x = 0$ to 3: graph goes up (increasing). $x = 3$ to 6: graph is flat (constant). $x = 6$ to 9: graph goes down (decreasing).

18 Reflections, rotations, and translations are rigid transformations — they preserve size and shape. Dilations and stretches change size.

19 Same shape but different size means the figures are similar (related by a dilation) but not congruent. Congruence requires equal size.

20 The rule $(x, y) \rightarrow (-y, x)$ is a 90° counterclockwise rotation around the origin.

21 Set up the proportion: $\frac{8}{x} = \frac{2}{5}$. Cross-multiply: $2x = 40$, so $x = 20$ cm.

22 $(2x + 10) + 3x + (x + 20) = 180$. Combine: $6x + 30 = 180$, $6x = 150$, $x = 25$.

Find more at
ViewMath.com/GA-Grade8

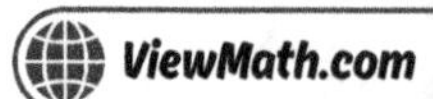

23. $9^2 + 40^2 = 81 + 1600 = 1681 = 41^2$. The converse of the Pythagorean Theorem confirms it's a right triangle.

24. $d = \sqrt{(4 - (-4))^2 + (3 - (-3))^2} = \sqrt{64 + 36} = \sqrt{100} = 10$.

25. $V = \frac{4}{3}\pi(3r)^3 = \frac{4}{3}\pi(27r^3) = 27 \times \frac{4}{3}\pi r^3$. The volume is 27 times as large.

26.   An association can be positive (both increase) and nonlinear (the trend is curved, not straight).

27.   A straight line of best fit only works for data with a linear pattern. Curved data needs a different model.

28. $y = -4(10) + 100 = -40 + 100 = 60$.

29.   The intersection of Girls row and Dog column is 12.

30.   Multiplying all values by 3 multiplies the mean by 3 and each deviation by 3, so MAD is multiplied by 3.

☑ Practice Test 4 — Answer Key

1 D	2 $\frac{53}{90}$	3 B	4 1.2	5 B	6 C	7 C	8 Taxi B	9 C
10 B	11 C	12 No	13 C	14 7	15 B	16 C	17 B	18 C	19 C
20 A	21 C	22 A	23 C	24 10	25 B	26 C	27 C	28 B

29.   Yes. Of instrument players, $\frac{15}{40} = 37.5\%$ play a sport. Of non-instrument players, $\frac{30}{40} = 75\%$ play a sport. The large differen

30.   C

Find more at
ViewMath.com/GA-Grade8

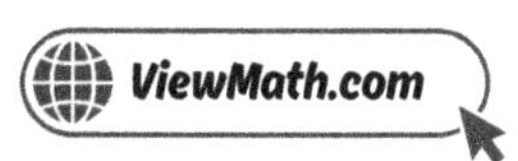

> 💡 **Time to Learn!** 💡
>
> *Review the explanations below, **especially for the questions you missed**.*
>
> *Understanding why each answer is correct builds stronger problem-solving skills.*
>
> ***Tip:*** *Circle any questions you got wrong, then read their explanation carefully.*

📖 Practice Test 4 — Detailed Explanations

1 $\sqrt{7}$ is irrational because 7 is not a perfect square. The other choices are all rational: $\frac{5}{8} = 0.625$, $0.\overline{6} = \frac{2}{3}$, and $\sqrt{9} = 3$.

2 Let $x = 0.5888\ldots$. Then $10x = 5.888\ldots$ and $100x = 58.888\ldots$. Subtract: $90x = 53$, so $x = \frac{53}{90}$.

3 $2\sqrt{5} = \sqrt{4} \cdot \sqrt{5} = \sqrt{4 \times 5} = \sqrt{20}$. Alternatively, $2\sqrt{5} \approx 4.47$ while $\sqrt{10} \approx 3.16$. They are not equal.

4 $5^0 = 1$ and $5^{-1} = \frac{1}{5} = 0.2$. So $1 + 0.2 = 1.2$.

5 $7 \times 7 = 49$, so $\sqrt{49} = 7$.

6 Move the decimal 3 places right: $6.1 \times 10^3 = 6{,}100$.

7 $\frac{2 \times 10^{-6}}{10^{-6}} = 2$. The bacterium is 2 micrometers long.

8 Taxi A: \$3/mile. Taxi B: $\frac{36}{9} = \$4$/mile. Taxi B is more expensive per mile.

9 $m = \frac{5-(-1)}{2-(-4)} = \frac{6}{6} = 1$.

Find more at
ViewMath.com/GA-Grade8

10. $-x + 5 = 2x - 1$. Add x: $5 = 3x - 1$. Add 1: $6 = 3x$, so $x = 2$. Then $y = -2 + 5 = 3$.

11. $60t + 80t = 420$, so $140t = 420$, $t = 3$ hours.

12. Input 7 maps to two different outputs (3 and 9), which violates the definition of a function.

13. Find the column where $f(x) = 15$. That column has $x = 4$.

14. The initial value is the output when $x = 0$, which is 7.

15. $7 - 3 = 4$, $11 - 7 = 4$, $15 - 11 = 4$. The constant change of 4 proves this is linear.

16. Slope $= \frac{4-(-2)}{4-0} = \frac{6}{4} = \frac{3}{2}$.

17. A downward straight line means the quantity decreases at a constant rate — water is leaking steadily.

18. Reflections are rigid transformations that preserve side lengths, angles, area, and perimeter.

19. Both rectangles have dimensions 6×10. A rotation maps one onto the other, so they are congruent.

20. $90°$ CCW: $(x, y) \rightarrow (-y, x)$. So $(0, -5) \rightarrow (5, 0)$.

21. Check ratios: $\frac{6}{3} = 2$ but $\frac{9}{5} = 1.8$. Since $2 \neq 1.8$, the sides are not proportional and the rectangles are not similar.

22. An exterior angle and its adjacent interior angle are supplementary: $180 - 140 = 40°$.

23. The ladder is the hypotenuse: $c^2 = 12^2 + 9^2 = 144 + 81 = 225$, so $c = 15$ ft.

  Find more at
ViewMath.com/GA-Grade8

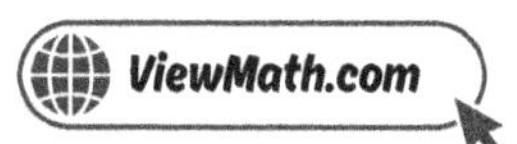

24 $d = \sqrt{6^2 + 8^2} = \sqrt{36 + 64} = \sqrt{100} = 10.$

25 $V = \frac{4}{3}\pi r^3 = \frac{4}{3}(3.14)(27) = \frac{4}{3}(84.78) = 113.04\ cm^3.$

26 A curved pattern is a nonlinear association. The data has a trend, but it is not a straight line.

27 $y = 1.5(4) + 2 = 6 + 2 = 8.$

28 $x = 50$ is far outside the data range (1 to 10), so this is extrapolation.

29 Students who do NOT play an instrument are much more likely to play a sport (75% vs. 37.5%). This difference in conditional percentages signals an association.

30 $|20 - 30| = 10$, $|40 - 30| = 10$. Both 20 and 40 have the greatest deviation (10). Among the choices, 20 is listed.

✅ Practice Test 5 — Answer Key

1 Rational   **2** B   **3** A   **4** B   **5** C   **6** 11 places to the left   **7** A   **8** D

9 B   **10** $(2.5, 3)$ **11** 14 and 36   **12** C   **13** 3   **14** C   **15** C   **16** $25   **17** C

18 C   **19** C   **20** B

21 $\frac{3}{2}$ **22** B   **23** B   **24** B   **25** B   **26** C   **27** Slope $= -1$; y-intercept $= 12$ **28** A

29 Totals: Grade A $= 30$, Not A $= 30$, HW Yes $= 40$, HW No $= 20$, Grand Total $= 60$. Relative frequencies: $\frac{24}{60} = 0.$

30 A

Find more at
ViewMath.com/GA-Grade8

💡 **Time to Learn!** 💡

*Review the explanations below, **especially for the questions you missed**.*

Understanding why each answer is correct builds stronger problem-solving skills.

Tip: *Circle any questions you got wrong, then read their explanation carefully.*

📖 Practice Test 5 — Detailed Explanations

1. $\sqrt{144} = 12$ exactly, because $12^2 = 144$. Since 12 is an integer, $\sqrt{144}$ is rational.

2. The repeating block has 2 digits, so multiply by $10^2 = 100$. Then $100x = 63.6363\ldots$ and $x = 0.6363\ldots$. Subtracting: $100x - x = 63$.

3. Perimeter $= 4 \times \sqrt{50} = 4\sqrt{50} \approx 4 \times 7.07 = 28.28 \approx 28.3$ cm. Note that $\sqrt{200} \approx 14.14$, which is only $2\sqrt{50}$ — that would be two sides, not four.

4. The star is at $\frac{1}{8}$. Since $\frac{1}{8} = \frac{1}{2^3} = 2^{-3}$, the answer is 2^{-3}.

5. $81 = 9^2$, so 81 is a perfect square.

6. The exponent is -11, so you move the decimal 11 places to the left from 2.5.

7. Same exponent, so add the coefficients: $5.2 + 3.8 = 9.0$. Answer: 9×10^6.

8. If the relationship passes through $(1, 5)$ and $(3, 15)$, then $k = 5$. But $5 \times 2 = 10 \neq 13$, so $(2, 13)$ does not fit.

9. $m = \frac{7-3}{\frac{3}{2}-\frac{1}{2}} = \frac{4}{1} = 4$

Find more at
ViewMath.com/GA-Grade8

10 Subtract the second from the first: $4y = 12$, so $y = 3$. Then $2x + 3 = 8$, $2x = 5$, $x = 2.5$.

11 Let smaller $= s$, larger $= l$. $s + l = 50$ and $2l - 3s = 30$. From first: $l = 50 - s$. Substitute: $2(50 - s) - 3s = 30$, so $100 - 5s = 30$, $5s = 70$, $s = 14$. Then $l = 36$.

12 The vertical line test checks whether any vertical line crosses the graph more than once. If it does, the graph is not a function.

13 $f(2) - f(1) = 7 - 4 = 3$.

14 $\frac{9-1}{2-0} = \frac{8}{2} = 4$. The rate of change is 4.

15 Nonlinear functions do NOT have a constant rate of change, which means their graphs are curved rather than straight.

16 $y = 5(3) + 10 = 15 + 10 = 25$.

17 A function can decrease while staying positive (e.g., dropping from 10 to 5). "Decreasing" means the values get smaller, not that they are negative.

18 A rotation is a rigid transformation that preserves side lengths, angle measures, and perimeter. Only the position and orientation change.

19 Applying the 90° CCW rule $(x, y) \to (-y, x)$: $R(0,0) \to (0,0)$, $S(4,0) \to (0,4)$, $T(0,3) \to (-3,0)$. This matches $R'S'T'$.

20 180° rotation: $(a, b) \to (-a, -b)$. Reflect over x-axis: $(-a, -b) \to (-a, b)$.

21 $\frac{9}{6} = \frac{3}{2}$, $\frac{12}{8} = \frac{3}{2}$, $\frac{15}{10} = \frac{3}{2}$. The scale factor is $\frac{3}{2}$.

Find more at
ViewMath.com/GA-Grade8

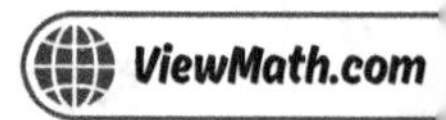

22   Alternate interior angles formed by parallel lines and a transversal are equal: $72°$.

23 $c^2 = 8^2 + 15^2 = 64 + 225 = 289$, so $c = \sqrt{289} = 17$.

24 $d = \sqrt{3^2 + 4^2} = \sqrt{9 + 16} = \sqrt{25} = 5$.

25 $V = \frac{1}{3}\pi r^2 h = \frac{1}{3}(3.14)(36)(10) = \frac{1}{3}(1130.4) = 376.8 \ cm^3$.

26   The independent (explanatory) variable goes on the x-axis; the dependent (response) variable goes on the y-axis.

27 $m = \frac{4-10}{8-2} = \frac{-6}{6} = -1$. $10 = -1(2) + b$, so $b = 12$.

28 $35 = 4x + 7$, $4x = 28$, $x = 7$.

29   Add rows and columns for totals. Divide each cell by 60 for relative frequency.

30   Lower MAD means less spread and more consistency. Class X (MAD = 5) is more consistent.

📋 Practice Test 6 — Answer Key

1 Rational   **2** C   **3** D   **4** C   **5** A   **6** B   **7** A   **8** B   **9** A

10 C   **11** B   **12** C   **13** 18   **14** B   **15** C   **16** A

17 Decreasing (fast at first, then slower), approaching constant.   **18** $(5, -3)$ **19** B   **20** $(-3, 4)$

21 A   **22** $56°$ **23** B   **24** B   **25** A   **26** B   **27** A   **28** B   **29** B   **30** B

Find more at
ViewMath.com/GA-Grade8

> ## 💡 Time to Learn! 💡
>
> *Review the explanations below, **especially for the questions you missed**.*
>
> *Understanding why each answer is correct builds stronger problem-solving skills.*
>
> ***Tip:*** *Circle any questions you got wrong, then read their explanation carefully.*

📖 Practice Test 6 — Detailed Explanations

1. $\frac{5}{6}$ is a fraction of two integers with a nonzero denominator, so it is rational. Its decimal is $0.8\overline{3}$, which repeats.

2. The repeating block 123 has 3 digits, so you multiply by $10^3 = 1000$.

3. $\pi \approx 3.1416$, $\sqrt{10} \approx 3.1623$, and 3.2 is exact. From least to greatest: $3.1416 < 3.1623 < 3.2$.

4. Any nonzero number raised to the zero power equals 1. So $(-2)^0 = 1$.

5. Edge $= \sqrt[3]{343} = 7$ in., since $7^3 = 343$.

6. Move the decimal 5 places left: $9.03 \times 10^{-5} = 0.0000903$.

7. $4 \times (9.5 \times 10^{12}) = 38 \times 10^{12} = 3.8 \times 10^{13}$ km.

8. $k = \frac{y}{x} = \frac{9}{2} = 4.5$.

9. $m = \frac{1-7}{5-2} = \frac{-6}{3} = -2$.

10. The solution to a system is the point (x, y) that satisfies both equations, which is where the lines intersect.

Find more at
ViewMath.com/GA-Grade8

11. $8 = 2 + 1.50m$. Subtract 2: $6 = 1.50m$. Divide: $m = 4$ movies.

12. Each input $(1, 2, 3, 4)$ has exactly one arrow, so each input maps to one output. Repeated outputs are allowed in a function.

13. $f(3) = 4(3) + 1 = 13$. $f(1) = 4(1) + 1 = 5$. Sum: $13 + 5 = 18$.

14. Function A's slope is -1 and Function B's slope is -3. Since $|-3| > |-1|$, Function B decreases faster.

15. $y = -2x + 6$ fits the form $y = mx + b$ with $m = -2$ and $b = 6$. A negative slope does not make a function nonlinear.

16. Slope: $\frac{16-7}{6-3} = \frac{9}{3} = 3$. Using $(3, 7)$: $7 = 3(3) + b \Rightarrow 7 = 9 + b \Rightarrow b = -2$.

17. The coffee cools quickly at first (steep decrease), then more slowly as it nears room temperature. The graph curves and levels off — nonlinear decreasing approaching constant.

18. The $90°$ CCW rule is $(x, y) \rightarrow (-y, x)$. So $(-3, -5) \rightarrow (5, -3)$.

19. Equal angles make triangles similar, but not necessarily congruent — their sides could be different lengths.

20. Reflecting over the x-axis keeps x and negates y: $(-3, -4) \rightarrow (-3, 4)$.

21. Congruent figures are a special case of similar figures with $k = 1$. All congruent figures are similar, but not all similar figures are congruent.

22. Corresponding angles formed by parallel lines and a transversal are equal.

23. $c^2 = 1^2 + 1^2 = 2$, so $c = \sqrt{2}$.

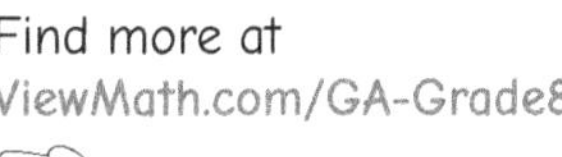
Find more at
ViewMath.com/GA-Grade8

24 The points are on the x-axis, so the distance is just the horizontal difference: $|5 - 0| = 5$.

25 Radius $= 12$. $V = \frac{4}{3}(3.14)(12^3) = \frac{4}{3}(3.14)(1728) = \frac{4}{3}(5425.92) \approx 7234.6 \; cm^3$.

26 As x increases, y increases. The dots go up-right, indicating a positive association.

27 $m = \frac{8-3}{5-1} = \frac{5}{4}$.

28 Extrapolation means using the model to predict values outside the data range, which can be unreliable.

29 The 20 is the count for a specific combination of two variables (7th grade AND soccer), making it a joint frequency.

30 MAD measures the average of the absolute deviations from the mean, describing how spread out data values are.

✅ Practice Test 7 — Answer Key

1 $\sqrt{2}$ (or $\sqrt{3}$) **2** C **3** ≈ 4.4 **4** C **5** 6 cm **6** B **7** 317 times **8** C

9 B **10** B **11** Length $= 16$ m, width $= 8$ m **12** C **13** B **14** C **15** Linear

16 B **17** C **18** A **19** C **20** $(-2, 5)$ **21** C **22** B **23** 29 **24** C

25 4 ft **26** C **27** $y = -1.5x + 9.5$ (approximately) **28** B **29** B **30** B

Find more at
ViewMath.com/GA-Grade8

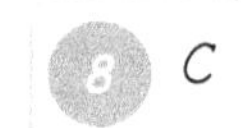

💡 *Time to Learn!* 💡

*Review the explanations below, **especially for the questions you missed**.*

Understanding why each answer is correct builds stronger problem-solving skills.

Tip: *Circle any questions you got wrong, then read their explanation carefully.*

📖 Practice Test 7 — Detailed Explanations

1. $\sqrt{2} \approx 1.414$ *is between 1 and 2 and is irrational because 2 is not a perfect square.* $\sqrt{3} \approx 1.732$ *also works.*

2. $\frac{36}{100} = 0.36$ *(terminates). The correct conversion uses* $99x = 36$, *giving* $\frac{36}{99} = \frac{4}{11}$. *The student treated* $0.\overline{36}$ *as if it were* 0.36.

3. $\sqrt{3} \approx 1.732$ *and* $\sqrt{7} \approx 2.646$. *Adding:* $1.732 + 2.646 = 4.378 \approx 4.4$.

4. *Numerator:* $9^3 \cdot 9^2 = 9^5$. *Then* $\frac{9^5}{9^4} = 9^{5-4} = 9^1 = 9$.

5. *Edge* $= \sqrt[3]{216} = 6$ *cm, since* $6 \times 6 \times 6 = 216$.

6. *Move the decimal 5 places left:* $350{,}000 = 3.5 \times 10^5$. *In scientific notation,* a *must satisfy* $1 \le a < 10$.

7. $\frac{1.9 \times 10^{27}}{6 \times 10^{24}} = \frac{1.9}{6} \times 10^3 \approx 0.317 \times 10^3 = 317$.

8. *In choice C,* $\frac{9}{3} = 3$ *but* $\frac{15}{6} = 2.5$. *The ratio is not constant, so it is not proportional.*

9. $m = \frac{12-3}{4-1} = \frac{9}{3} = 3$.

Find more at
ViewMath.com/GA-Grade8

10 $-2x + 10 = x + 1$. So $9 = 3x$, $x = 3$. Then $y = 3 + 1 = 4$. Solution: $(3, 4)$.

11 $2l + 2w = 48$ and $l = 2w$. Substitute: $2(2w) + 2w = 48$, $6w = 48$, $w = 8$. Then $l = 16$.

12 In a function, the first coordinate x is the input and the second coordinate y is the output.

13 $f(8) = \frac{8}{2} + 6 = 4 + 6 = 10$.

14 Both slopes are 5, so the rate of change is the same. The y-intercepts are 3 and -7, which are different.

15 It fits $y = mx + b$ with $m = 8$ and $b = 1$. The variable has exponent 1, so it's linear.

16 Slope: $\frac{11-5}{2-0} = \frac{6}{2} = 3$. The y-intercept from $(0, 5)$ is $b = 5$. So $y = 3x + 5$.

17 An upward graph means the output (distance) is getting larger as the input (time) increases — the object is moving away.

18 Add the translation to each coordinate: $(1 + 3, 2 + 1) = (4, 3)$.

19 Congruent figures must have equal corresponding side lengths. $4 \neq 6$, so the squares are not congruent (though they are similar).

20 Multiply each coordinate by $\frac{1}{2}$: $(-4 \times \frac{1}{2}, 10 \times \frac{1}{2}) = (-2, 5)$.

21 All squares have four $90°$ angles. Any two squares can be mapped by a dilation since the ratio of their sides is constant. So all squares are similar.

22 One angle is $90°$ and another is $32°$. Third angle: $180 - 90 - 32 = 58°$.

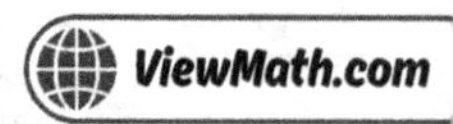

23 $c^2 = 20^2 + 21^2 = 400 + 441 = 841$, so $c = \sqrt{841} = 29$.

24 The right triangle has legs 8 and 6. Distance $= \sqrt{8^2 + 6^2} = \sqrt{64 + 36} = \sqrt{100} = 10$.

25 $\frac{4}{3}\pi r^3 = \frac{256}{3}\pi$. Divide by π: $\frac{4}{3}r^3 = \frac{256}{3}$. Multiply by $\frac{3}{4}$: $r^3 = 64$. So $r = 4$ ft.

26 A strong positive trend corresponds to an r-value close to 1. $r = 0.9$ indicates a strong positive association.

27 Using $(1, 8)$ and $(5, 2)$: slope $= \frac{2-8}{5-1} = \frac{-6}{4} = -1.5$. $8 = -1.5(1) + b$, $b = 9.5$. Line: $y = -1.5x + 9.5$.

28 In $y = mx + b$, the slope m represents the rate of change: how much y changes for each 1-unit increase in x.

29 Of 42 pizza lovers, 24 are boys. $\frac{24}{42} \approx 0.571$.

30 Deviations: $2, 2, 0, 4, 4$. MAD $= \frac{2+2+0+4+4}{5} = \frac{12}{5} = 2.4$.

✓ Practice Test 8 — Answer Key

1 C **2** $\frac{81}{99} = \frac{9}{11}$ **3** ≈ 8.7 feet **4** B **5** -5 **6** C **7** B **8** A **9** B

10 $(4, 5)$ **11** B **12** B **13** 5

14 Function A: rate $= 3$, initial $= 20$. Function B: rate $= 5$, initial $= 5$. At $x = 10$, $B = 55$ is greater than $A = 50$.

15 Nonlinear, each output is x^3. **16** 7 **17** The speed is constant for 4 seconds. **18** C

19 Reflection over the y-axis **20** C **21** 80 **22** A **23** C **24** $\sqrt{13}$ **25** 300 cm^3

 **26** C    **27** Approximately 2 to 2.25    **28** C

 **29** Boys: $\frac{20}{30} \approx 66.7\%$. Girls: $\frac{12}{20} = 60\%$. The percentages are close, so association is weak or none.    **30** B

💡 **Time to Learn!** 💡

Review the explanations below, **especially for the questions you missed**.

Understanding why each answer is correct builds stronger problem-solving skills.

Tip: Circle any questions you got wrong, then read their explanation carefully.

 **📖 Practice Test 8 — Detailed Explanations** 

1 Every integer n can be written as $\frac{n}{1}$, making it rational. Not every square root is irrational (e.g., $\sqrt{4} = 2$).

2 Following the same pattern: $99x = 81$, so $x = \frac{81}{99}$. Simplify by dividing both by 9: $\frac{81}{99} = \frac{9}{11}$.

3 Side $= \sqrt{75}$. $8.6^2 = 73.96$ and $8.7^2 = 75.69$. Since 75 is between these, $\sqrt{75} \approx 8.7$.

4 By the negative exponent rule, $\frac{1}{a^n} = a^{-n}$, so $\frac{1}{2^4} = 2^{-4}$.

5 $(-5)^3 = -125$, so $\sqrt[3]{-125} = -5$.

6 $1 \times 10^2 = 100$ kg. From the chart, the lion's bar reaches close to the 10^2 mark, making its mass closest to 100 kg.

7 $\frac{6.4 \times 10^{10}}{4 \times 10^6} = 1.6 \times 10^4 = 16{,}000$ photos.

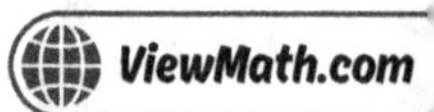

8.   Machine A: 12 bottles/hour. Machine B: $\frac{50}{5} = 10$ bottles/hour. Machine A is faster.

9. $\frac{50-68}{6} = \frac{-18}{6} = -3°F$ per hour. The negative sign shows the temperature is decreasing.

10.   Subtract: $y = 5$. Then $x + 5 = 9$, so $x = 4$.

11.   At month 2, X is cheaper ($\$60 < \$65$). At month 3, Y is cheaper ($\$80 < \$85$). They cross between months 2 and 3.

12.   Each student has exactly one student ID, so each input (name) maps to exactly one output (ID). That makes it a function.

13. $-3x + 15 = 0 \Rightarrow -3x = -15 \Rightarrow x = 5$.

14.   A: slope 3, $b = 20 \Rightarrow A(10) = 3(10) + 20 = 50$. B: slope 5, $b = 5 \Rightarrow B(10) = 5(10) + 5 = 55$. Function B is greater at $x = 10$.

15. $1^3 = 1$, $2^3 = 8$, $3^3 = 27$, $4^3 = 64$. The differences $(7, 19, 37)$ are not constant, confirming nonlinear.

16. $1 = -3(2) + b \Rightarrow 1 = -6 + b \Rightarrow b = 7$.

17.   A flat section means the output (speed) stays the same — the object travels at a steady speed for those 4 seconds.

18.   A $180°$ rotation uses the rule $(x, y) \to (-x, -y)$. So $(-4, 3) \to (4, -3)$.

19.   Each x-coordinate is negated while y stays the same: $(1, 1) \to (-1, 1)$, etc. This is a reflection over the y-axis.

20.   Dilation by factor 2: $C(3, 3) \to (2 \times 3, 2 \times 3) = (6, 6)$.

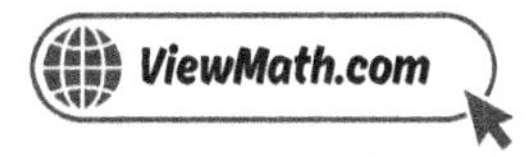

21. *New side:* $5 \times 4 = 20$. *Perimeter:* $4 \times 20 = 80$.

22. $180 - 48 - 67 = 65°$.

23. $a^2 = 13^2 - 5^2 = 169 - 25 = 144$, *so* $a = 12$.

24. $d = \sqrt{2^2 + 3^2} = \sqrt{4 + 9} = \sqrt{13}$.

25. *A cylinder is 3 times the volume of a cone with the same base and height:* $100 \times 3 = 300 \ cm^3$.

26. *When data points group closely together, it is called a cluster.*

27. *Using endpoints:* $m \approx \frac{11-2}{5-1} = \frac{9}{4} = 2.25$. *A slope around 2 is a reasonable estimate.*

28. *For each 1-unit increase in* x, y *increases by the slope (6). For 3 units:* $6 \times 3 = 18$.

29. *Compare the conditional relative frequencies. Since 66.7% and 60% are similar, the preference doesn't change much by gender.*

30. *MAD* $= 0$ *means every deviation is 0, so every data value equals the mean.*

📋 Practice Test 9 — Answer Key

 C  B  B  A  B  B  C  10 *packages*  A

 $(2,5)$  B  B  B  A  B  A  A  C  C

 C  B  B  B  B  C  A  A  B  C

  Find more at
ViewMath.com/GA-Grade8

 $MAD = 6$

💡 **Time to Learn!** 💡

Review the explanations below, **especially for the questions you missed**.

Understanding why each answer is correct builds stronger problem-solving skills.

Tip: Circle any questions you got wrong, then read their explanation carefully.

 **Practice Test 9 — Detailed Explanations**

1. $1.41421356\ldots$ is the decimal expansion of $\sqrt{2}$, which is non-repeating and non-terminating, indicating an irrational number.

2. Let $x = 0.444\ldots$. Then $10x = 4.444\ldots$. Subtract: $9x = 4$, so $x = \frac{4}{9}$. Note that $\frac{4}{10} = 0.4$ (terminates), which is different from $0.\overline{4}$.

3. $\sqrt{2} \approx 1.414$, so $3\sqrt{2} \approx 4.243$. This is between 4 and 5.

4. $3^{-2} = \frac{1}{3^2} = \frac{1}{9}$.

5. $7^2 = 49$ and $8^2 = 64$. Since $49 < 50 < 64$, we have $7 < \sqrt{50} < 8$.

6. $8 \times 10^5 = 800{,}000$ and $3 \times 10^6 = 3{,}000{,}000$. The higher exponent wins: 3×10^6 is larger.

7. $7 \times 8 = 56$ and $10^{-3} \times 10^{-2} = 10^{-5}$. Then $56 \times 10^{-5} = 5.6 \times 10^{-4}$.

8. Service P: $4 \times 10 = 40$ packages. Service Q: $k = \frac{10}{2} = 5$, so $5 \times 10 = 50$ packages. Difference: $50 - 40 = 10$.

Find more at
ViewMath.com/GA-Grade8

9 $m = \frac{7-(-2)}{3-0} = \frac{9}{3} = 3.$

10 Both tables give $y = 5$ when $x = 2$. The solution is $(2, 5)$.

11 $c + w = 20$ and $2c + 4w = 56$. From first: $c = 20 - w$. Substitute: $2(20 - w) + 4w = 56$, so $40 + 2w = 56$, $2w = 16$, $w = 8$.

12 In choice B every input (3, 4, 5) maps to exactly one output. The other sets repeat an input with different outputs.

13 $f(x)$ is function notation. $f(3)$ means substitute 3 for x in the function rule — it is NOT f times 3.

14 The initial value is the y-intercept (b). Function A has $b = 8$ and Function B has $b = 2$. Since $8 > 2$, Function A starts higher.

15 A straight line means constant rate of change, which is the definition of linear. Negative slope means the line goes down, but it is still linear.

16 Slope: $\frac{-1-(-4)}{1-0} = 3.$ y-intercept: -4 (from $x = 0$). So $y = 3x - 4$.

17 The ball goes up, reaches a peak, then comes down. Gravity causes the rate of change to vary, so the path is curved (nonlinear).

18 A translation moves the figure to a new position. Side lengths, angle measures, and parallelism are all preserved.

19 A dilation by a factor other than 1 changes the size of the figure, so it does not preserve congruence.

20 Reflect over x: $(x, y) \rightarrow (x, -y)$. Then reflect over y: $(x, -y) \rightarrow (-x, -y)$. The combined result $(x, y) \rightarrow (-x, -y)$ is a $180°$ rotation.

Find more at
ViewMath.com/GA-Grade8

21 Divide the image coordinates by the original: $\frac{12}{4} = 3$ and $\frac{-6}{-2} = 3$. The scale factor is 3.

22 The three interior angles of any triangle always add to $180°$.

23 $5^2 + 12^2 = 25 + 144 = 169 = 13^2$. This is a Pythagorean triple.

24 The distance formula uses the horizontal and vertical distances as legs of a right triangle and applies $a^2 + b^2 = c^2$.

25 $V = \pi r^2 h = 3.14 \times 9 \times 10 = 282.6 \; cm^3$.

26 Clusters are groups of points that bunch together, while outliers are isolated far from the trend. Both can appear in the same scatter plot.

27 The line goes from about $(0.5, 6.5)$ to $(7.5, 1.5)$. Slope $\approx \frac{1.5 - 6.5}{7.5 - 0.5} = \frac{-5}{7} \approx -0.71$.

28 $y = 5(8) + 20 = 40 + 20 = 60$ dollars.

29 Two-way tables deal with categorical data (counts and proportions). Mean scores involve numerical data, not categorical.

30 Mean $= 100$. Deviations: $0, 5, 5, 10, 10$. MAD $= \frac{0 + 5 + 5 + 10 + 10}{5} = \frac{30}{5} = 6$.

📋 Practice Test 10 — Answer Key

 B  D  B  C  C  Cell C, Cell A, Cell B 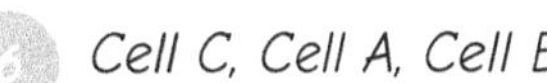 3×10^6  99

 B  B  7 T-shirts and 8 hats  C  9  B  B  $y = 6x - 3$

 Find more at
ViewMath.com/GA-Grade8

 ViewMath.com

 B    $(1,4)$ 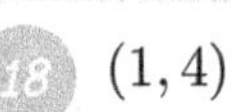 B    A    C    B    13 ft    A    B

 Nonlinear — height increases in childhood, then levels off in adulthood.    C    C    A

 A

💡 Time to Learn! 💡

Review the explanations below, **especially for the questions you missed**.

Understanding why each answer is correct builds stronger problem-solving skills.

Tip: Circle any questions you got wrong, then read their explanation carefully.

📖 Practice Test 10 — Detailed Explanations

1. 5 is not a perfect square, so $\sqrt{5}$ is irrational and belongs in the Irrational region.

2. In Step 5, the student divided 2 by 10 instead of 9. From $9x = 2$, the correct answer is $x = \frac{2}{9}$.

3. $\pi \approx 3.14$, so $\pi^2 \approx 3.14 \times 3.14 = 9.8596 \approx 9.9$.

4. A negative exponent flips the fraction: $\left(\frac{3}{4}\right)^{-2} = \left(\frac{4}{3}\right)^2 = \frac{16}{9}$.

5. $\sqrt{2}$ cannot be expressed as a fraction of two integers, so it is irrational.

6. Cell A: 8×10^{-3} mm. Cell B: 5×10^{-2} mm. Cell C: 3×10^{-4} mm. From smallest to largest: $3 \times 10^{-4} < 8 \times 10^{-3} < 5 \times 10^{-2}$, so Cell C, Cell A, Cell B.

7. $\frac{3.6}{1.2} = 3$ and $10^{10-4} = 10^6$. Answer: 3×10^6.

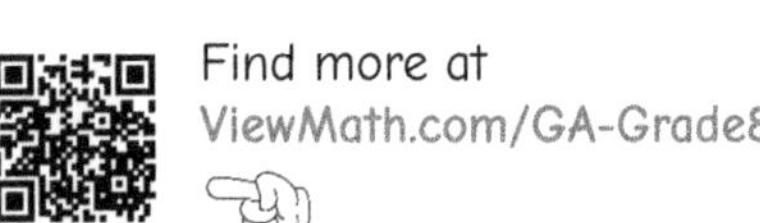

8 $k = \frac{54}{6} = 9$. When $x = 11$: $y = 9 \times 11 = 99$.

9 A line that goes down from left to right has a negative slope.

10 Add: $2x = 14$, so $x = 7$. Then $y = 10 - 7 = 3$. Solution: $(7, 3)$.

11 $t + h = 15$ and $12t + 8h = 148$. From first: $h = 15 - t$. Substitute: $12t + 8(15 - t) = 148$, $4t + 120 = 148$, $4t = 28$, $t = 7$. Then $h = 8$.

12 Any circle fails the vertical line test because a vertical line will cross it at two points. Therefore a circle is never a function.

13 From the graph, $f(1) = 3$ and $f(4) = 6$. So $f(1) + f(4) = 3 + 6 = 9$.

14 $G(4) = 14$. $H(4) = 4(4) + 1 = 17$. Since $17 > 14$, Function H has the greater value at $x = 4$.

15 A linear function is written as $y = mx + b$, where m is the slope and b is the y-intercept.

16 Substitute $m = 6$ and $b = -3$ into $y = mx + b$: $y = 6x - 3$.

17 Standing still means distance doesn't change: horizontal line. Then walking at a steady pace means distance increases at a constant rate: line sloping upward.

18 The $90°$ CCW rule is $(x, y) \to (-y, x)$. So $(4, -1) \to (1, 4)$.

19 When triangles are congruent, all corresponding sides are equal. Since AB corresponds to DE, $DE = 8$ cm.

20 A $180°$ rotation uses $(x, y) \to (-x, -y)$. So $(2, -7) \to (-2, 7)$.

21 Similar figures can be mapped by a combination of rigid transformations (translations, reflections, rotations) and dilations.

22 $180 - 54 - 54 = 72°$.

23 $d = \sqrt{12^2 + 5^2} = \sqrt{144 + 25} = \sqrt{169} = 13$ ft.

24 $AB = \sqrt{(6-1)^2 + (13-1)^2} = \sqrt{25 + 144} = \sqrt{169} = 13$. The others: $AC = 12$, $AD = 12$, $BD = \sqrt{49 + 144} = \sqrt{193}$.

25 $200\pi = \pi r^2 (8)$, so $r^2 = \frac{200}{8} = 25$, and $r = 5$ cm.

26 Height rises quickly in youth, slows, and plateaus in adulthood. This is a nonlinear pattern.

27 A good line of best fit has roughly half the points above and half below.

28 $y = 2(4) + 1 = 9$.

29 $\frac{24}{80} = 0.30$.

30 Set A is spread from 3 to 11 (range 8). Set B is tightly clustered around 7 (range 2). Set A has more spread.

Well done checking your answers!

Keep practicing to strengthen your skills.

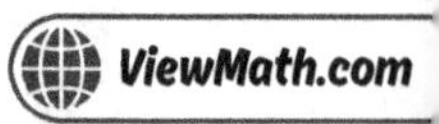